COSMOLOGY

or

Universal Science

(1888)

Containing the Mysteries of the Universe Regarding God, Nature, Man, the Macrocosm and Microcosm, Eternity and Time, Explained According to the Religion of Christ by Means of the Secret Symbols of the Rosicrucians of the Sixteenth and Seventeenth Centuries, translated from an old German manuscript, and provided with a Dictionary of Occult Terms

Franz Hartmann

ISBN 0-7661-0730-2

Request our FREE CATALOG of over 1,000
Rare Esoteric Books
Unavailable Elsewhere

Freemasonry * Akashic * Alchemy * Alternative Health * Ancient Civilizations * Anthroposophy * Astral * Astrology * Astronomy * Aura * Bacon, Francis * Bible Study * Blavatsky * Boehme * Cabalah * Cartomancy * Chakras * Clairvoyance * Comparative Religions * Divination * Druids * Eastern Thought * Egyptology * Esoterism * Essenes * Etheric * Extrasensory Perception * Gnosis * Gnosticism * Golden Dawn * Great White Brotherhood * Hermetics * Kabalah * Karma * Knights Templar * Kundalini * Magic * Meditation * Mediumship * Mesmerism * Metaphysics * Mithraism * Mystery Schools * Mysticism * Mythology * Numerology * Occultism * Palmistry * Pantheism * Paracelsus * Parapsychology * Philosophy * Plotinus * Prosperity & Success * Psychokinesis * Psychology * Pyramids * Qabalah * Reincarnation * Rosicrucian * Sacred Geometry * Secret Rituals * Secret Societies * Spiritism * Symbolism * Tarot * Telepathy * Theosophy * Transcendentalism * Upanishads * Vedanta * Wisdom * Yoga * *Plus Much More!*

KESSINGER PUBLISHING, LLC
http://www.kessingerpub.com
email: books@kessingerpub.com

COSMOLOGY,

OR

Cabala.   Universal Science.   Alchemy.

CONTAINING

THE MYSTERIES OF THE UNIVERSE,

REGARDING

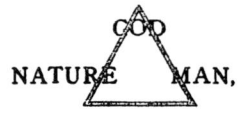

NATURE   MAN,

THE

Macrocosm and Microcosm,

ETERNITY and TIME

EXPLAINED ACCORDING TO

THE RELIGION OF CHRIST,

BY MEANS OF

THE SECRET SYMBOLS

OF THE

ROSICRUCIANS

OF THE SIXTEENTH AND SEVENTEENTH CENTURIES.

COPIED AND TRANSLATED FROM AN OLD GERMAN MANUSCRIPT, AND PROVIDED WITH A DICTIONARY OF OCCULT TERMS

BY

FRANZ HARTMANN, M.D.

BOSTON:
OCCULT PUBLISHING COMPANY,
120 TREMONT STREET.
1888.

N.2 J. sayo

This Book is in great part a translation of a german work published in Altona in 1785-1788. The same plates are to be found in the german book. The introduction and the glossary have been added

German title.
"Geheime Figuren aus den Rosenkreuzern

PART I.

AUREUM SECULUM REDIVIVUM

or

The Ancient Golden Age,

which has disappeared from the Earth, but will reappear;

whose germ is beginning to sprout, and will bear blossom and fruit.

by

HENRICUS MADATHANUS THEOSOPHUS,

Medicus & tandem, Dei gratia, aureæ crucis frater.

Translated from the German.

"If there is one among You who is deficient in wisdom, let him pray to the spirit of truth, who comes to the simple-minded, but does not obtrude upon any one, and he will surely obtain it."—*Jacob Epist. v. 5.*

SYMBOLUM AUTHORIS:

Centrum Mundi; Granum Fundi.

THE SECRET SYMBOLS OF THE ROSICRUCIANS.

INTRODUCTION

(BY THE TRANSLATOR).

A FEW centuries ago the name "Rosicrucian" produced a great stir in the world. It suddenly and mysteriously appeared on the mental horizon, and as mysteriously disappeared again. The Rosicrucians were said to be a secret society of men possessing superhuman — if not supernatural — powers; they were said to be able to prophesy future events, to penetrate into the deepest mysteries of nature, to transform Iron, Copper, Lead, or Mercury into Gold, to prepare an *Elixir of Life* or *Universal Panacea*, by the use of which they could preserve their youth and manhood; and moreover it was believed that they could command the *Elemental Spirits of Nature* and knew the secret of the *Philosopher's Stone*, a substance which rendered him who possessed it all-powerful, immortal, and supremely wise.

Many historical facts seem to confirm the truth of such statements, and certain still-existing legal documents go to prove that gold on certain occasions has been indeed produced by artificial means; but the Rosicrucians always insisted that this art was only one of the most insignificant parts of their divine science, and that they possessed far more important secrets. Some of those people believed to be Rosicrucians could heal the sick by the mere touch of their hands, or by means of some wonderful medicines, and they performed some extraordinary feats which equalled those recorded in the Christian Bible and in other sacred books and histories of ancient religions. Some were believed to have attained an age of several hundred years; some, are believed to be still living upon this earth. The Rosicrucians themselves did not contradict such stories; on the contrary, they asserted that there were many occult laws and mysterious powers, of which mankind on the whole knew very little at those times, and which would for many centuries to come remain unknown to "science"; because all science is based upon the observation of facts, and facts must be perceived before they can be observed; but the spiritual powers of perception are not yet sufficiently among mankind as a whole to enable them to perceive spiritual things. They say that if our spiritual powers of perception were fully developed, we should see this universe peopled with other beings than ourselves, and of whose existence we know nothing at present. They say that we should then see this universe filled with things of life, whose beauty and sublimity surpass the most exalted imagination of man, and we should learn mysteries in comparison with which the art of making gold sinks into insignificance and becomes comparatively worthless. They speak of the inhabitants of the four kingdoms of nature, — of Nymphs, Undines, Gnomes, Sylphs, Salamanders, and Fairies, — as if they were people with whom they were most intimately acquainted, and as if they did not belong to the realm of the fable, but were living beings of an ethereal organization, too subtle to be perceived by our gross material senses; but living, conscious, and knowing, ready to serve and instruct man and to be instructed by him. They speak of *Planetary Spirits* who were formerly men, but who are now as far above human beings as the latter are above animals, and they seriously assert that if men knew the divine powers which are dormant in their constitution, and were to pay attention to their development, instead of wasting all their life and energies upon the comparatively insignificant and trifling affairs of their short and transient external existence upon this earth, they might in time become like those planetary spirits or gods.

We are not in a position to demonstrate to what extent such assertions made by ancient and modern Rosicrucians are true, or whether those accounts have been exaggerated or misunderstood, nor would we expect to be believed if we were to put forward our testimony to strengthen a doctrine rejected by modern scientific authorities who have never seen anything but what can be seen by means of the external senses. We do not desire to dispute with those who are incapable of seeing in man more than an intellectual animal, who are extremely skeptical in regard to the existence of an invisible world within the visible one, but who are vain and credulous enough to believe that nothing can possibly exist of whose existence they know nothing; and that if anything spiritual or divine were to exist, in spite of their assertions to the contrary, they would have found it out long ago. We have no desire to quarrel with the learned about such matters; because the existence of the Unseen cannot be proved so long as it is invisible for them, and even the existence of the sun remains a mere matter of opinion and speculation for those who are blind.

What can a purely material science know about Spirit or about God? what can a science which deals merely with the details of the external phenomena of life know about the fundamental, invisible principles which are the external causes of the universal manifestations of life?

There were true and false "Rosicrucians" during the Middle Ages, as there are true and merely nominal "Christians" to-day. The Pseudo-Rosicrucians were very numerous; the true ones were seldom to be found. Some people believed to be Rosicrucians were imprisoned in dungeons and tortured, with a view to extract their secrets from them; but nothing was gained by such persecutions, because divine things cannot be revealed to him who has not the capacity to comprehend such revelations. No one can be taught how to employ spiritual powers which he does not possess, and no one possesses spiritual powers unless he becomes spiritual himself. No one can be taught to be a good artist or musician unless he possesses a natural talent for the exercise of such arts; no one can be taught how to exercise spiritual functions unless he possesses the organs required for them. As well might we attempt to instruct an animal how to use human speech, as to attempt to teach an unspiritual person to exercise spiritual powers and to become an Alchemist. Such attempts would always end with a failure; because the laws of nature are unchangeable, and no being can enter a higher state than that to which its nature is adapted. Intellectuality is not identical with Spirituality, but merely a product of spiritual activity in its incipient stage; only when man has outgrown his animality can his organization become a fit instrument for the exercise of divine powers and a proper temple for the habitation of God.

Although the ancient Rosicrucians were visible men, inhabiting mortal and visible bodies, nevertheless they were highly spiritually developed beings, in whom the occult powers, dormant in the constitution of all men, had become unfolded to such an extent that they could control the action of the universal principle of Life, and obtain power over certain secret forces in nature; and they were therefore able to perform deeds which must necessarily appear incredible or miraculous to those who do not possess such powers. This ignorance of the secret forces of nature is the cause why all modern scientific and historical researches regarding the true nature of the Rosicrucians have been a failure; and their character and history is not understood merely because the true character and nature of that being which we call *Man* is not understood, nor his full history known.

What and who were the Rosicrucians? The question is answered by the echo: What and who is Man? So long as we know nothing of man, except his external anatomy and physiology, we cannot hope to be able to judge about the sources of his emotional and intellectual functions, much less about the divine attributes which the real inner man — the regenerated spirit — possesses. If we want to know anything about the divine inner man, the consciousness of our own divinity must first become alive within ourselves, and we must attain self-knowledge; for man cannot actually *know* anything except that which exists within himself — all other learning is merely speculation, guess-work, belief, and opinion.

A little reflection will prove the truth of this statement: If we look at any external object, say — for instance — a tree, we perceive nothing of it except the image which its reflection creates in our mind, consequently within our own selves. How, then, could we know anything about a thing which does not exist in our own mind, but in the mind of another. It is true that another person may give us a description which will enable us to form an image in our mind resembling, to a certain extent, the image in the mind of the other; but such an image is our own product; it is merely our own creation which we have created with the help of another, and we therefore know nothing but that which we have ourselves created. Moreover, if we see, feel, smell, hear, or taste a thing, we know, for all that, nothing about it; all we know are the sensations which it produces on our organism, and if our organization were different, the sensations received would be different. We therefore know nothing about the thing itself, but merely our *relations* to it. How, then, could we know anything about a thing to which we stand in no relation, or in a relation of which we are unconscious? This is an old philosophical doctrine, which is theoretically accepted by our scientists and philosophers, but which they are continually disregarding in practical life, because they have not yet fully awakened to a realization of its truth.

What we know about eternal things is therefore merely the relation in which we stand to their external *appearance*, while of the invisible powers, which are the causes of such external appearances, we know absolutely nothing; because they produce no impressions upon our minds, and are therefore non-existent within our own selves. It is true that we may employ our fallible intellectual powers and draw logical deductions in regard to the unknown, by reasoning from the basis of that which *we imagine to know:* but this is not true knowledge; it is merely speculation, theory, and opinion. Such theories and opinions may be true or false; they may be good enough until new discoveries are made, which overthrow them, and upon which new theories are built up, to be overthrown in time again by others. This is not the kind of knowledge upon which spiritual science is based. Real knowledge is the result of a direct perception and understanding of the truth; only when the truth exists within ourselves can we know it; and we can know it only by the knowledge of self.

The only things which modern science actually knows is the external nature of things as they appear; but there are certain powers latent within the constitution of man, which, if they become developed, call a higher scale of internal senses into activity, which may enable him to receive spiritual impressions, and to hear, see, feel, taste, and smell things which far surpass the powers of perception of the external senses; and as the latter may be educated by use, likewise the former may be made more acute and receptive by practice.

All men possess this power of interior perception, to a certain extent; he who would deny this fact would deny his own reason; for "Reason" is the spiritual or intuitional perception of a truth; it is "*Common Sense*," whose decisions are frequently contradicting the logic of the calculating intellect. This power of *Intuition*, or, as we would define it, the *Feeling of a truth*, is, in the majority of men, merely in a rudimentary state,—an uncertain thing, a sensation easily overruled by the speculating intellect; but in him whose spirit has awakened to a consciousness of his divine existence, its light grows bright and its voice becomes strong, and it calls into life the inner senses by which man may see and perceive the beings and things existing in the realm of the Soul of the Universe and the inner causes of all external phenomena, and behold the beauties of a spiritual existence of which material science dares not even begin to dream.

Who can imagine or describe the glories and beauties of the Unseen? Living in a world of gross material forms, we know nothing about the ethereal forms of Life which inhabit the immensity of space; we are prone to imagine that we know all that exists, but our reflection tells us that the infinite realm of the Unknown is as much greater than the realm of that which is known as the ocean is greater than a pebble lying upon its shore. Nature is one great living whole, and the spiritual power acting within her is omnipotent and eternal. He who desires to know Universal Nature and the Eternal Spirit, must rise above personal and temporal considerations, and look upon nature from the standpoint of the Eternal and Infinite. He must, so to say, step out of the shell of his limited and circumscribed personal consciousness, and rise up to the top of the mountain, from which he may enjoy a view of the wide expanse of the All. He who lives at the periphery sees only a part of the All; only from the centre of the circle can we survey the actions of light in all its directions as the beams radiate from the centre. Therefore, the Rosicrucians say that he who knows the *One* knows All, while he who believes to know many things, knows only the illusions of the shadow produced by the light of the One.

The small cannot embrace the great, the finite cannot conceive of the infinite; if men desire to know that which is immensely superior to their personal selves, they must step out of those selves and by the power of Love embrace the infinite All.

How many who crave for occult knowledge are willing to renounce that personal self, which is so dear to them, and around whose existence are centred all their hopes, cares, and affections? How many of those who desire to be instructed in occult science are willing to accept and to realize practically the truth of the first doctrines of Occultism; namely, that the Universal Spirit is One, and that in him and by his power we live and have our being, and that we should love Wisdom above all, and all humanity,—yea, all living beings,—as if they were parts of ourselves? Are not such and similar truths proclaimed every day from all pulpits in Christian and heathen countries, and are they understood, realized, and practically followed out by the hearers or by the preachers themselves? Or are they mere words, impressed upon the memory, listened to by the ear, but neither coming from the heart nor penetrating to it? Verily, if those truths were realized and practised, the *Golden Age* would soon again appear upon the earth, and we should meet angels and saints, Adepts and Rosicrucians, at every step.

This renunciation of one's own beloved personal self, with all its desires, theories, and intellectual speculations, is the great stumbling-block in the way of the searchers after the truth, barring the way to the entrance of the light at the threshold of the soul. It is "the stone which the builders rejected, and which has become the head of the corner. Whosoever shall fall upon that stone shall be broken, but on whomsoever it shall fall, it will crush him." It is the one unavoidable and necessary condition for those who desire to obtain eternal life; for how could they partake of the consciousness of the Universal Spirit so long as they cling to the consciousness of being merely a very limited personality?

Upon the recognition of this truth are based all the fundamental doctrines of the religions of the world; it is the rock (*Petra*) upon which the universal spiritual church of humanity is built; it is allegorically represented in the *Bhagavad-Gita* by the battle which *Arjuna* has to fight with his own personal *Egos*, to enable him to become united to *Krishna*; it is represented by the Christian *Cross* adorned with the figure of a dying man; for it is not the Christ-principle which dies upon a cross, but the semi-animal self which must suffer and die so that the real man may rise into a glorious resurrection, and become united with the light of the *Logos*,—the *Christ*. It is not physical death which is represented in this beautiful allegory, but the *mystic death*, the death of personal desires, personal claims, and personal considerations. Physical death is a matter of little importance, so far as the spirit of man is concerned: it is merely one of the many similar incidents which man has to experience during his eternal career; and physical man dies, is born, and dies again many times before he reaches that state in which he needs no more to be born and to die. The mystic death refers to the cessation of man's existence as a separate and isolated being and his elevation into a higher state, preparatory to his entering into *Nirvana*.

To grasp this sublime idea, it will, above all, be necessary to form a correct conception of the true nature of Man. It is acknowledged by all, except the most superficial observers, that the external form of man, whose anatomy we know, is not the real thinking and feeling inner man, but merely an external expression of the latter. What else can this inner man be but an invisible power, active within the physical form? This internal power, called the Spirit of Man, has established a centre of life in the heart and a centre of thought in the brain; it sends the blood from the heart to all parts of the physical organism, and the light emanating from the brain radiates along the nerves and communicates thoughts to the most distant organs of sense. Unconsciously, but nevertheless effectively, the soul acts in the workshop called a human being, guiding the processes of life and building up a form in which the character of the spirit becomes expressed in each part of the external shape.

Man leads three different kinds of existence. Two of these states are known to all; the third is known only to those who possess the power of spiritual perception, and for all others it is merely a matter of speculation. The first state in which man exists as a personal human being is as a child in the womb of its mother. There he leads an almost merely negative existence, knowing nothing at all about the existence of the outer world, with its inhabitants, its life, light, and sound. Entombed in the womb of his mother he has nothing else to do but to grow. Even if he were able to think and to reason, a state of existence outside of that womb would be incomprehensible to him, because it is beyond his experience; and we might easily imagine a body of scientists in the fœtal state holding a meeting, and by drawing logical deductions from what they know, proving scientifically and satisfactorily to each other that any other existence but that within the womb is a scientific impossibility, and a belief in it a deplorable delusion. At last, however, the great moment arrives; in spite of all scientific reasoning the child is born, and enters into a new, and at first incomprehensible, existence. It is now surrounded by light and sound, which begin to attract its attention. Things which in its former state were of supreme importance for its welfare,—such as the *placenta*, the liquor *amnii*, the *umbilical cord*, etc.,—are now of no importance whatever, and have become perfectly worthless. The new man begins to grow; he sees other beings beside his own self, which, like himself, seem to have a life of their own; he feels himself bodily separated from other forms; he feels bodily wants, pleasures, and pains, which are not shared by others; and thus the illusion of self is created, and that self appears to be of supreme importance to him. All of man's thoughts, desires, and aspirations are now centred around that personal self. He studies how he may increase its pleasures and comforts, how he may keep it from suffering and prolong its existence. That which concerns his own self appears to him to be the only thing needful; that which concerns others, as a matter of secondary consideration, because he feels, knows, and enjoys only the existence of his own self.

Many human beings die before they have seen the light of the terrestrial world, or soon after they are born; many human beings die before they have gotten over the delusion of self, and awakened to a higher state of existence: comparatively few are born into the light of the eternal life in the spirit, by the process of *spiritual regeneration*. This spiritual state is as far superior to man's terrestrial existence as the latter is to his fœtal state; and yet it is unknown to science and incomprehensible to the superficial reasoner. We cannot *know* what it is, so long as we have not experienced it; but we may, even by logical reasoning, convince ourselves that such a state exists.

If we study the processes by which the existence of external things is brought to our inner consciousness, we easily understand that the mind of man is not a thing enclosed within the narrow limits of the physical man; but that while the consciousness of man is centred within his organization, the substance of mind must necessarily reach as far as man's thoughts can reach. Occult science teaches that the spiritual power which constitutes the real man, and whose centre of activity is in the heart of man, whence it radiates to all parts of his organism, is a universal principle which fills, surrounds, and penetrates all things. Likewise the influence of the rays of the physical sun is manifest everywhere, penetrating into the seeds and germs of plants, and developing their forms according to their individual characters. The sun, without leaving his place in the sky, acts by the influence of his power within the forms of terrestrial things, causing a tree to grow out of a kernel in which no such tree could possibly have been contained. Likewise the universal eternal power of the spiritual Sun of the Universe enters the heart of man, and may develop an immortal being.

A ray of spiritual light enters the heart and stimulates the higher elements of the soul into activity and life. It establishes — so to say — a centre of polarity in the soul, causing the spiritual germ to expand and live a higher life than that of which the physical man is conscious; to breathe a spiritual ether, too subtle to support the life of the animal form, and to obtain a knowledge of spiritual truths, far surpassing the conception of mortals. The powers of the terrestrial sun enter the heart of a tree and cause the growth of branches and twigs, the development of flowers and fruits; they live even in the invisible odor which emanates from the trees, and which may perhaps be perceived even before we approach the latter. The powers of the celestial sun of grace enter the heart of man, and cause the development of a soul whose activity extends far beyond the limits of the physical body.

Life, being a function of the eternal Spirit, causes the development of body and soul. It enables the physical organism to assume a shape resembling that of its parents, and adapted to the conditions in which it is destined to live; but when the physical form has attained its full development, spiritual activity does not cease. The physical body of man may have attained its apex of growth, and its strength may begin to decline, and yet man may grow stronger in love and stronger in knowledge, and acquire more wisdom even during old age. It moreover seems that the development of the higher spiritual faculties is facilitated when the animal energies begin to decline; because the power which in a former state of growth was used to promote the development of the body can then be employed for the unfoldment of the soul. All this goes to prove that man's visible body is not the real man, but that the latter is an invisible power, which may grow even during terrestrial life into a being of great magnitude, while only the kernel — the physical body — is visible to the imperfect sensuous perceptions of mortals.

This Light, being the Life and the Truth shining into the hearts of men, is the *Christ*, or Redeemer of mankind. It is universal, and there is no other redeemer; it is known to the wise of all nations, although they do not all call it by the same name; it existed in the beginning of creation, and will exist at its end; it is the flesh and blood, the substance and power, of the inner spiritual man in his highest divine aspect.

For all we know, the inner man lives in his house — called the physical body — merely during the time when the latter is in a state of wakefulness and conscious of its external surroundings. When the external form is asleep, the inner man may be fully awake and live in a higher state, far more appropriate to his nature and dignity; but when the physical man awakens again, it may remember nothing about the experiences of the spiritual self, because the latter is far superior to the former, and has a memory of its own. These assertions are not a mere matter of speculation, but known to all who have investigated the dual nature of man; and, moreover, there are certain conditions under which the invisible man may manifest his powers and tell us of his experiences during the sleep of the visible form; and such conditions are met with in cases of trance, somnambulism, and ecstasy.

Universal science teaches that man's spiritual and invisible self is a being far superior to man's visible and personal self, and that the former does not fully enter the latter, but may be looked upon as its guardian spirit, overshadowing it with his wings. This spiritual self lived before the physical body was born, and will continue to live when the latter is again dissolved in the elements; it may have overshadowed many other personalities before it gave light and life to its present external expression; it may have inhabited many a house of flesh and blood, and taken from each its most precious jewels to ornament itself.

Such, indeed, is the ancient doctrine of reincarnation. It is taught by the religions of old, and was known to the Rosicrucians of the Middle Ages. It teaches that only the higher self of man is immortal, and that he who desires to enter the eternal Life must strive to grow out of his lower animal self, and become able to unite his soul with his own spiritual *Ego*, or "Christ." He who succeeds in accomplishing this during his terrestrial life may even now share the superior life and the attributes of that higher existence, in which he may alternately become one with the supreme source of all Good, from which his spirit emanated in the beginning of time.

The Christian teachings, as well as the Brahminical books, whose origin dates from prehistorical times, all tell the same tale in various allegorical forms. They all say that original man, a pure spiritual being, emanated in the beginning from the eternal substance of Parabrahm. This celestial *Adam* was the *Christ* or the *Word*, existing with God and being God itself from all eternity. "In him was life, and the life was the light of men." This spiritual *Man* was an expression of the Will and the Thought of God, and could therefore have no other thought and no other will but that of his eternal source. He was bisexual; that is to say, in him existed the male and female elements in a state of harmony before they became, in consequence of his differentiation in material forms, to a certain extent separated from each other. Gradually this divine Man was tempted by the illusions of his senses, which became more misleading as his organization grew to be more material. He began to think and to will in a manner deviating from the will and imagination of God; he "ate from the tree of knowledge" — he became material and sank into matter. The original spiritual power, constituting aboriginal Man, became differentiated into terrestrial men and women, embodied in material forms, subject to the sufferings caused by the influences of the elements and exposed to the vicissitudes of terrestrial life; and it is now the supreme object of man's existence upon this earth, by living up to the universal law, to subject and purify the animal elements existing in his constitution, to assume his former spiritual state of purity, and by bringing his thoughts and actions in perfect harmony with the will and imagination of God, to become reunited with the light of the *Logos*.

This fundamental truth forms the laws of all true religion, and all the principal religious systems upon this globe are founded upon this final unification with God. The wise men of all ages know of the birth of Christ, not of a man called "Christ," but of the divine Saviour, who may be born in every human heart. The Christ is the "Son of God," a ray of Light from the eternal spiritual sun of the universe, shining into the hearts of men, and growing up in the midst of the semi-material elements of man's organization. Nature produces the Christ. She is an eternal mother, for all forms are evolved from Nature, and all return again into her womb. Yet she is an ever *immaculate virgin;* for she has no connection with any external God; the fructifying power of the "*Holy Ghost*" lives and acts within her own centre.

These truths are as old as the world, and they had been known many thousands of years before the advent of modern Christianity. They have often been impressed upon mankind by great reformers and sages, and have been again forgotten by man. It is said that at certain periods, when mankind as a whole begins to forget the old truths, when religions become materialized by a forgetting of the secret signification of its symbols, and by a blind *belief* in external forms, taking the ascendancy over the true spiritual power of *Faith*, a new *Avatar*, or "Christ," appears upon this earth to refresh men's memory, and to teach anew the old truths by his word and by his example. For all we know, *Jesus of Nazareth* may have been such a reformer, penetrated by the light of the *Logos*. If the accounts given about him are true, he was a man filled with the divine spirit of Christ, and therefore a Christ himself, as every other man may be, if he succeeds in entering the light of the Christ. He was a man penetrated by divine power, a man in whose soul Divinity took a form, and therefore he was a god; and likewise every other man in whom the same power grows into a form may be a god like Christ.

The duality of man in his material and spiritual aspect is universally recognized; it is a truth which forces itself continually upon the attention of every one who is able to think. The final union of the divine elements existing in the organization of man, with the sum and substance of the divine elements existing in nature, is the truth upon which all reasonable religious teachings are based. All the principal religious systems take, however, only the two extreme views of man's existence into consideration; namely, the existence of man in a supreme spiritual state, being one and identical with God, as a drop of water is one and identical with the the ocean, — although even in that state the drop, the individual man, is conscious of existence in the whole, and enjoying a happiness of which we can form no conception: the other extreme is that of man's existence upon this earth as a material corporeal being; that is to say, as a spirit, bound to an earth-formed material body which hampers the free movements of the spirit, and whose sensations, desires, and temptations continually tend to drag the real man still deeper down into animality and materiality.

But between these two extremes there are innumerable other forms of existence, in which man may lead a conscious life, free from the bonds of gross matter, living in a comparatively ethereal form, under higher and superior conditions than those to which he has been accustomed during his life upon the earth; in a world where his thoughts assume for him an objective reality, and where he enjoys happiness or suffers evil according to the nature of the spiritual forces which he has set in motion during his earthly career. As life exists in thousands of manifold forms upon this earth, likewise there may be thousands of manifold modes of existence in the spiritual state; as the number of suns appearing on the sky during a cloudless night is so great that, looked at from the standpoint of the Infinite, our little globe appears like an insignificant speck of dust in the universe of suns and revolving worlds; likewise the states of existence which man in a higher state may enter must be as numerous as his states upon this earth, and some of them may be so far superior to our present existence that the latter may indeed be looked upon as an exile or punishment for our *sins;* in other words, as the effect of evil *Karma* created by ourselves during some previous existence upon this earth. While terrestrial life lasts at best a hundred years, our life in the spiritual state may last for thousands of years before, by the law of *Karma* (the law of cause and effect on the moral plane), we shall again be forced to overshadow a newborn human being and to submit to a new incarnation.

Man is a *dual* being in one of his aspects; but in another aspect his nature is *triune*, while looked at from other points of view he appears under still more varied aspects. In his triune aspect he appears not merely as a trinity of Spirit, Soul, and Body, but this trinity exists on three different planes. The highest plane is the divine state in which man lives in the light of the Logos, and the light of the Logos in him. The lowest state is his semi-animal existence, in which he may be conscious of the light merely by feeling its presence in moments of deep meditation. Between these two states man exists as a spiritual, but not yet purely divine, power, leading a life as much superior to terrestrial man as the life of the latter is superior to that of a plant; and yet they have not yet entered or attained the highest state, in which all consciousness of separation and isolation ceases to exist, and where man becomes one with God.

The doctrines of the ancient Hermetic philosophers, and more recently the theories presented by Darwin, go to show how the universal principle of Life acting within primordial Matter continually evolves new and higher forms, so that, to speak in the language of the Rosicrucians, in the course of millions of ages, "a stone becomes a plant, a plant an animal, an animal a man, and a man a god." Everywhere, in the mineral, vegetable, and animal kingdom, we see innumerable gradations of existence, without any hard lines of separation between them; or, if such lines are seen, it is because the "missing links" have been lost. Moreover, there are amphibious beings which are equally adapted to live in the air and in the water, in the earth and air, or in the earth and the water; and the same may be said about the Elementals, or Spirits of Nature.

As there are innumerable gradations of visible forms, likewise there are innumerable gradations of invisible ones; and there are amphibious existences in the realm of the Universal Soul which may exist in two different states, — which may appear sometimes in visible forms, while at other times they are invisible to us. There are beings on the *Astral Plane* which are only seen by those who have developed their inner senses to an extent sufficient to enable them to perceive such forms; but certain conditions may exist which may cause such ethereal forms to become more dense and material, so as to become perceptible even to the physical senses of man. The soul of man is such an amphibious being.

His corporeal, external form is visible, his spirit is invisible to the eye; but the soul, which connects spirit and body, may live in or out of the body; and while as a rule it is invisible to us, still its ethereal shape may, under certain conditions, become dense enough to be visible, and even tangible, to mortals.

To fully grasp this idea, it will be necessary that we should rise above the opinions and prejudices which a reliance upon the impressions produced by the illusions of the senses have created within the sphere of our mind. We should keep sight of the fact that the whole of the universe, with all its forms, is nothing else but materialized and visible thought; that is to say, images which existed subjectively within the imagination of the Great First Cause, and have been thrown into objectivity by its own will. If our organization were sufficiently ethereal, we should, perhaps, not be aware of the existence of gross matter; but we would see the souls of men, animals, plants, and minerals, and we should be able to see a thought as soon as it had become formed in the mind of a human being.

The majority of people whose minds have not become perverted by the insane doctrines of a false and superficial science, or by the superstitious doctrines presented by modern priestcraft under the guise of religion, believe that the above-described spiritual condition of man is one into which they will enter after the hand of death has dissolved the bonds which chain the spirit to its earthly tabernacle; but occult science proves that such condition may be arrived at even during terrestrial life. Moreover, if we desire to continue to exist in a high spiritual state after the death of the physical body, we should attempt to enter that state even during our terrestrial life; for "death" is merely a throwing off of that which has become useless: it can work no miracles, nor endow us with any attributes which we have not legitimately attained by our own evolution while in the physical state.

The true *Brothers of the Golden and Rosy Cross* were men who had attained that state. Their existence is as well proved as any other fact in history; but their true nature can be known only to him who has become like them. Such persons have existed at various times, and they exist now; we may read of such persons in the sacred books of the East, in the lives of the Christian saints, and in mediæval and modern occult literature. Being spiritual themselves, those Adepts are able to see and to converse with other spiritual beings, being as much acquainted with the world of invisible causes as with the world of external effects: they are able to guide and control the processes of life by which matter is changed and transformed; having gained spiritual life, they have found the *Universal Panacea* which cures all ills appertaining to a lower state of existence; standing upon the rock of the true living *Faith* — meaning the direct spiritual perception and knowledge of causes and effects — they have come into the possession of the *Philosopher's Stone*, which lifts them above the region of doubt, and establishes within their own hearts the *Cube of Life*, the corner-stone of the temple of Wisdom.

What is to hinder a man who has acquired the consciousness and knowledge of being a spiritual power, independent of a material organization, to leave his house of flesh and blood at will whenever he chooses, and to re-enter it again whenever it is necessary to do so? Consciousness is realization of a state of existence, and man cannot become conscious of being that which he is not; he must first become spiritual before he can know that he is a spiritual being: to imagine to be something which one is not would merely be an idle assumption. If man has once become spiritual, he will perceive the existence of other spiritual beings like himself, and of those which are inferior to him. He will then be comparable to an amphibious being, — able to live upon the earth, or to dive into the eternal ocean of Spirit.

Such beings are still men, and it must not be supposed that even in their spiritual state they have neither form nor organization, or that they were, like a breath of air, without form and consciousness. No thought is possible without an organized mind, no function can be exercised without organs of some kind; form is an external expression of internal attributes, and wherever exists a character, there exists a form in which it finds its expression. The superior character of spiritual beings finds its expression in divine and ethereal forms: Will, Imagination, and Expression in Form are the trinity which forms the basis of all existence in the visible and invisible universe.

Such beings, while in their spiritual state, are independent of the conditions of space and time as we conceive of them; "matter" to them is not impenetrable: they can see into the hearts of men and read their innermost thoughts; in their human state they are like other men, and subject to human conditions. Thus they lead two different kinds of existence, — as men upon the earth and as angels in heaven; and when death destroys their physical bodies, it will be a matter of little importance to them, for it will merely destroy that which they do not need, and which they have learned to do without and to care no more for than a man would care for his warm winter-coat when the summer appears. This divine state may, perhaps, be attained by all after untold ages, and in the slow progress of human evolution; but to enter it now requires efforts, and all efforts, to be effective, must be well directed and based upon a true knowledge of the nature, the origin, and final destiny of man. Theory without practice avails little; but to make practice available, it should be preceded by a correct theory, the true religion, the science of Universal Life.

How can we expect that a man grown up in the midst of scientific misrepresentations and erroneous theological dogmas, fed with prejudices and superstitions, should be able to comprehend and realize such exalted truths? True religion and vile priestcraft are so inextricably mixed up together, at present, that it is almost impossible for average man to distinguish between the two. If the truths of Christianity are taught, those who cannot discriminate between the true and the false will accept at once the superstitions which clerical ignorance and assumption have mixed with the former, or they will reject the true with the false. If the idolatry and superstitions of modern churchmen are exposed, those who cannot think deeply will reject all religion, and instead of seeing the truth, they will sink still deeper into the mire of materialism. Animals cannot be reasoned with; they can only be ruled by love or by fear; but the great majority of men are still too much surrounded by their own animal elements to be able to see the truth even if it is shown to them; and until they become amenable to reason, the existence of priestcraft will be a necessary evil to them. But when will priests and clergymen awaken to a comprehension of the mysteries of the religion which they teach, and begin to show the people the truth, instead of clamoring for a belief in fables and of a literal acceptance of allegories? They will not arrive at that state until they, too, succeed in vanquishing the selfish propensities of their own animal natures, and rise up to the conception of the true and universal God, whose temple is man, and who needs no deputies or representatives upon the earth to make his will known to each individual man. They will not know the truth until they become true servants of the God of Humanity, instead of being merely servants of their churches, whose energies are employed to serve the temporal benefits of the latter.

The theory of modern Christianity is not in harmony with its practice: in our modern churches theory and practice contradict each other at every step. The true spiritual church of the living Christ is built upon the rock of the living Faith, a power by which spiritual truths are recognized; but the modern church is based upon popular ignorance regarding the laws of existence, and held together by selfish and personal considerations. According to the Bible, God is a universal Spirit, and can be approached only through the Light of the Christ; but modern church-practice makes of God the caricature of a man, and of the priest an unavoidable medium for communication with him. To the mind of the vulgar, a faith in God is something which is far beyond the powers of their comprehension, while a belief in the priest is of supreme importance; for the former is ever unapproachable, while the latter can be approached. Such misconceptions are suffered to continue, because they advance the temporal interests of the church. God is dethroned from his seat, and his place is occupied by the priest; and thus the "Beast of Babylon" will sit upon the throne until the strong arm of awakened reason will throw it down, and divine justice send it down to the "bottomless pit."

When man ceases to be a child and begins to think, the truth that to live cannot be the true object of life forces itself upon his mind. He asks the great question: "Who am I; and what is the object of my existence?" and he expects his parents or teachers to answer. To give the correct answer to these questions should be the principal object of religious or scientific education. How do modern science and religion answer him? The former teaches him all about his external form, its anatomy, physiology, and of his relations to his external surroundings; but this is not all the knowledge he wants. He knows that the external body will die, and to know the conditions of its short existence is merely a small and comparatively insignificant part of that great science which deals with the inner spiritual man, who may live for millions of years, or be forever immortal. Man wants to know something definite about his own spiritual self, and about the conditions under which that self may exist independent of its physical form. Science remains mute, or treats his questions with derision and contempt. He therefore turns to religion, and religion answers — but what is her answer?

Some two hundred and fifty Christian sects — with perhaps a greater number of not-Christian ones — tell him the most contradictory things, nearly all of their statements being opposed to each other. They give no positive proofs for their statements, nor any intelligible reasons for their beliefs; but each sect claims to possess the truth, and some of them go so far as to condemn to eternal punishment every one who does not accept their opinions. Their beliefs usually refer to some historical event, said to have taken place many centuries ago, and about which no reliable historical documents exist; or they refer to the interpretation of certain passages in some books, which may be interpreted in various ways. Thus the seeker after the truth who expects to obtain it by means of external information, is driven from one port to another, like a ship without a rudder blown about by contrary winds. There are many who at last stifle the voice of reason which speaks in their heart, accept some form of belief, and persuade themselves that thereby they will attain the salvation they coveted; many others, in whom the intellectual element is more dominant, conclude that the truth cannot be found, and must remain forever unknown. They cease to inquire, or they adopt the superstitions of superficial materialistic reasoners, and of those many henceforth cease to care for anything but for their own personal self and its temporal acquisitions.

But the true searcher after the truth will not remain satisfied so long as the unknown exists. Having examined the various altars and not found the true God, he at last approaches the altar of the unknown God, around which nothing but darkness exists. But in the centre of his own heart there burns a divine fire; and lighting the lamp of his intellect at this divine fire of Reason, he begins to see the truth, and finds it far more sublime than he ever dared to hope. Not in books or in religious doctrines of any kind can the truth be found, nor in intellectual speculations. If we desire to know the truth, we must permit it to enter our heart, so that it may become a part of ourselves. Then by the power of self-knowledge we may see the truth in its own light; we may feel it and see it and know what it is.

If we learn only one religious system and blindly accept its doctrines, we are easily misled into a belief that this system alone contains all the truth; but if by the light of reason we examine the various religious systems of the world, study their symbols and allegories and their secret meaning, comparing the various doctrines with each other, we will soon find that they all have one grand fundamental truth, of whose existence the greater number of the keepers of these religions are ignorant, or whose sublimity, even if they teach it, they do not realize; namely, the *One-ness* of Nature and the divine eternal Spirit within.

Oh, how far greater than the god of the churches is the God of the Universe! He is not a limited being to be coaxed and persuaded by priests, but an eternal power, unchangeable as the Law. The god of the churches is a bugbear whom the people cannot really love, because they are taught to be afraid of him; the God of humanity is the eternal power of Love, the source of all being, whose image exists in the heart of the pure, whose nature is Fire, whose rays are the Light of Intelligence and the principle of immortal Life. There are thousands of nominal Christians living a life of sensuousness and selfishness, hoping to be able to cheat the god of their church at the end of their days and to obtain salvation from him in spite of their sins; but the universal God creates within the heart of each man his own judge, who cannot be persuaded by words, but who reads the innermost thoughts and judges each man according to his true merits.

Universal religion is based upon the recognition of the truth that all humanity is one, and that we should always be guided in our actions by our considerations for the welfare of all, in preference to all personal considerations. Sectarianism, on the other hand, appeals to men's selfish desires, either in regard to this life or in regard to the dread hereafter. It teaches them that they should seek above all salvation for their own selves; the salvation of others is a matter of secondary consideration. The real sectarian bigot would be ready to annihilate the world, if he could thereby evade the death of his person or prolong the existence of the latter. Likewise the scientific bigot would be glad to destroy the truth, if he could thereby save from perdition the artificial system of theories which he has laboriously constructed. Modern science and modern religion teach that personal happiness, either in this or a future existence, is the great *desideratum;* occult science teaches that Humanity is a whole, that the personal man is merely a transient illusion, and that permanent happiness cannot be obtained until that illusion of self is destroyed.

The truth of the latter doctrine is self-evident, as may be seen, if we observe the daily life of man. The more a man is thinking of his own self, the more unhappy and discontented will he grow; the more he forgets his own self, the happier will he be. Why do the people run after pleasures and pastimes, why do they love intoxicating drinks, music and shows, theatrical performances and sensuous distractions of all kinds, if not because during such moments they are able to forget their own selves? Men do not become unconscious or die as soon as they cease to think of their own selves; but their souls expand on such occasions beyond the limits of the prison-house of material clay, and they enjoy for a short period a superior form of existence. Sensuous pleasures, however, convey no permanent happiness; they are not lasting, and are often followed by reactions which are injurious and become the causes of suffering.

Sensuous man lives in the impressions which external objects produce upon his senses; intellectual man lives in a world of thought of which his brain is the creator and which is real to him; spiritual man lives in a spiritual world of beauty which the divinity in his heart has created for him and which is the image of heaven. The animal is happy if its physical wants are supplied; for it knows no other but the sensuous world; and if everything in that world relating to the animal is in harmony, then that world is in harmony with itself. In merely intellectual man, filled with an erroneous conception of the importance of his own personality, the world of thought which he has created for himself is not in harmony with the outer world. Such a person has continually something to desire; he sees things, not as they are, but as he imagines them to be. In spiritual man the world of thought and the world of will are in exact harmony; he recognizes the truth and sees the things as they are, without personal consideration. Looking at the world, not from a personal standpoint and without considering himself as something separated and isolated from other beings, he recognizes the action of universal laws by the power of his enlarged perception.

Here we may be permitted to add a few words of explanation for those who are not sufficiently acquainted with the doctrines of occult science to understand what we mean. Life, Consciousness, and Sensation are not—as a superficial "science" is prone to believe—products of the physical organization of man; but they are states or functions of that universal principle which men call "God," and which become manifest in the forms of terrestrial beings. All the forms of Life in the Universe may be looked upon as being manifestations of the One and Universal Principle of Life in various forms; the whole of the Cosmos, being a product of the Universal Mind, may be regarded as universal, absolute consciousness becoming relative in separate forms. The universal consciousness of the Universal Mind forms spiritual centres of consciousness in living beings, whereby each being may feel and know its surroundings; and as the mind of living beings expands, their consciousness and power of sensation and perception expand with it; for all their powers belong to the mind, and not to the body: the latter without the mind is merely a form without life. Wherever a man's consciousness exists, there exists the real man. As long as man's consciousness is centred within the animal principles of his organization, he will merely be conscious of being an animal, and he will live in the sphere of his sensual emotions. If a man's consciousness is entirely centred in his brain, he will live in a world of speculation and ideas; if a man's consciousness becomes established within the divine elements of his soul, it will expand with his soul and lift him up into the higher regions of thought, where he becomes unconscious of being in a state of limitation and lives in a higher condition of existence, incomprehensible to those who have never experienced it. Such a state of existence, without consciousness of one's own personality, is described by the apostle Paul, who was "caught up" into such a state, and the same condition is known to the Indian saints, who give it the name *Samadhi*.

In the final renunciation of one's own personal Self consists the victory over death and the resurrection of the spirit; it is the *mystic death*, represented by the Christian *Cross*, a symbol known thousands of years before the advent of modern Christianity upon this earth. The symbol of the Cross is seen everywhere in Christian countries, upon the spires of the churches, in chapels and dwellings, and by the roadside; but to the great majority of priests and laymen it is nothing else but a memento, to call up the memory of an event said to have taken place nearly 2000 years ago in Palestine, on which occasion a perfect and divine man was executed like a criminal, falling a victim to the ignorance of the clergy and the vanity of the Pharisees of his time. A belief in the actuality of this occurrence, by which God is said to have become reconciled to man, is held among the Christians to be of supreme importance for one's future salvation; although no intelligible reason is given to show that God was ever angry with Man and that any such reconciliation was necessary; nor is it explained why a certain opinion in regard to an event of which we cannot possibly know by experience, should be necessary to attain the eternal life of the spirit. Those, however, whose eyes are not blinded by dogmas and who have compared the allegories of the Christian religion with the allegories of the Eastern religions know that—

whether the historical account of the crucifixion of Christ has actually taken place as recorded, or whether it is purely symbolical,— the symbol of the Cross has a far deeper and far more sublime secret signification. It represents an episode in the history of every one who has become a Christ; it is the symbol of spiritual regeneration, through which all have to pass who desire to enter into the divine state of existence. In the mind of the superficial thinker the Cross is a token of torture and death; in the conception of the enlightened it is a symbol of victory over self, of triumph, and the beginning of immortal life. Rivers of blood have been caused to flow and millions of human beings have been murdered in the attempts made by professed "Christians" to force all the nations of the world to adopt or profess a certain *opinion* — such as was authorized by the Roman Catholic Church — in regard to a certain historical event concerning the life and death of a man who taught that the fundamental doctrine of his religion was universal love and benevolence! To those murderers "in the name of Christ" the Cross was a symbol of pillage, destruction, and robbery, and there was no crime too villanous not to be permissible if perpetrated under the banner of the Cross. The Cross of religious bigotry ruled at the *autos da fé* of the church and sanctioned the burning of living victims; it filled the dungeons of the Holy Inquisition and inspired the villanies committed in the chambers of torture.

The Christ who lives in humanity shudders, if he thinks of all the horrible crimes which have been perpetrated in his name by those who misunderstood his true nature. Could such things have been possible if the "believers in Christ" had understood the true signification of the Cross, instead of merely clinging to the external form? Could sectarian intolerance exist to-day, if the real meaning of the Cross were understood by the priests? Could the Christ-spirit of the nineteenth century, the spirit of truth, have been insulted by the dogmas of Immaculate Conception and Papal Infallibility, if the self-styled deputies of Christ knew their own Master? All the religious bigotry, intolerance, superstition, and degradation in Christian countries finds its root and cause in a misunderstanding of the doctrines which they profess to believe. Our legally appointed guardians of the truth, like the Pharisees of old, do not know the truth; they talk about it, but their practice does not harmonize with their theories. The truth is so beautiful that he who has succeeded in knowing it will practise it and cling to it forever.

Let him who desires to feel and to know the true meaning of the Cross step out of the gloomy temples where terror and fear, ignorance and priestcraft, have established their throne, and let him worship the true living God, the light and Holy Ghost pervading all nature, the source of all life from man down to the insect, yea, even to the spark of life slumbering in a stone; the source of all glory and power, knowledge and wisdom, love and harmony; whose activity is manifest everywhere, and whose image should be seen in every human heart. Let him leave priests and monks to their psalmodies and to the contemplation of a dread hereafter, which they have often just cause to fear, and let him enter the Living Light which makes even the external material world resplendent with beauty. Let him step out of the musty libraries of our speculative and superficial science and study the book of nature in the light of the latter. Let him brush away the cobwebs which have accumulated in his own chamber; so that the light of truth may enter the windows of his soul and melt the icy crust around his heart and cause him to realize the sublimity and majesty of the God of both Christians and Heathens, the God of the Universe, whom no one can approach, but whose nature may be known in the manifestation of his power which evolved the Cosmos.

Who can know a thing which he has never seen? Who is legitimately entitled to speak as an authority about things which are beyond his own comprehension? Who can be the legitimate keeper of truths which are beyond his mental horizon and of which he knows nothing? Only he who has grown to be divine — not he who has merely assumed that title — can know divine things; only he who has stepped out of the bonds of matter and become spiritual can know the things of the spirit. Only he whose internal senses are open is able to see and converse with the beings of the super-terrestrial plane. Men talk and dogmatize about the attributes of God, as if they were well acquainted with him; they sermonize about Love and Charity, Faith and Wisdom, without understanding practically the meaning of these terms. Who can know what Love is but he who has loved? Who can know Wisdom but he who is wise? Who can know the Truth but he in whom the Truth dwelleth? Who can know God but he who has identified his own soul with him?

Man is originally a son of God. If he wants to know the *Father*, he must return to his original divine state and become a *Christ*, full of the *Holy Ghost*, the Light of the *Logos*. He is a child of eternally immaculate Nature; if he wants to know his mother, he must enter into perfect harmony with her and become natural. How can man know Nature as she is so long as he is himself unnatural and imagines her to be otherwise than she is? How can he understand Nature so long as he does not let her light enter his heart, but looks merely at his own unnatural misconceptions regarding her and which he has himself created in his mind? Before man can develop any spiritual powers he must first re-establish harmonious relations between himself and universal Nature; only when he has become natural can he expect to grow spiritual and to be able to obtain command over the divine powers of his mother True natural science is therefore the basis of all true religion; but to obtain a true knowledge of Nature we must study her as she is, not as she has been represented by those who are continually misrepresenting her, and who know nothing about her except some of her external forms.

To know Nature as she is, and not as she is supposed to be by others, we must free our mind of all the prejudices and misconceptions which have become established therein by a merely superficial science and by a dogmatic theology based upon an entire misconception of the true nature of man. We must free our mind from the noxious influences arising from the animal elements existing within our own soul, so that our understanding will become clear and the light of truth may shine through the pure atmosphere of our own internal heaven without any clouds obstructing its way. We must become One with Nature and One with the Truth, and by the knowledge of Self we will then know the Truth and have the powers of Nature at our command.

It is one of the fundamental truths of occult science that individual man is an image of Nature. His constitution is based upon the same laws upon which Nature as a whole is constructed, and as a child resembles its mother; likewise man's organism resembles universal nature in everything but the external form. He is a Microcosm of the Macrocosm of nature; containing within himself, either germinally, potentially, or actively, all the powers and principles, substances and forces contained within the great organism of nature, and moreover the great and the little world continually act and react upon each other; the elementary forces of nature act upon man, and the forces emanating from man — even his thoughts — react again upon nature; and the more harmony there exists between man and the laws of universal nature, the more intimate will be the connection between the two: for the two are actually only *one*, the fact of their appearing to be two being merely an illusion which has been caused by man's contravention of the laws of Nature, and by his consequent falling into an unnatural state. Let man again become a true child of Nature, and of *one mind* with his mother, and he will know all nature by knowing himself. He will then be like the *lost son* mentioned in the Bible, who returns again to the house of his father and has his natural birthright and inheritance restored to him. Let him establish the throne whereon the truth may reside within his own heart, and he will know the truth without the study of books and without theoretical speculation.

In the *Secret Symbols of the Rosicrucians* the science of Nature as a whole, with all the invisible powers living and acting therein, have been laid down, as it appeared to those great and wise men who were in harmony with Nature and able to read her in her own light. Those symbols are easily comprehended by him who finds the key to their understanding within his own heart; but to all others they will be unintelligible, because they will see merely their external forms and cannot enter their spirit. As the body of a man is a thing without life, after the life-principle which acted therein has deserted it, likewise these symbols of old are things without life unless they become again alive within our own selves. Modern misconceptions and misinterpretations have built a wall around the Temple of Truth, through which we must break before we can enter the latter; but if an opening has once been made in the wall through which the light may shine, then those symbols will serve us as guides and help us to understand the truths which we feel and see within our own selves; and without such an interior perception true understanding is impossible, and we must remain in the field of mere theory and speculation.

Our present age is eminently given to intellectual and theoretical speculation; while even the meaning of the term "*Wisdom*," *i.e.* that knowledge which one can only find at the bottom of one's own heart, has been lost. In our age things are done in haste and in hurry, and they are done in a very superficial manner if religious truths are concerned. We now travel at the rate of sixty miles an hour to a distance which it took our ancestors weeks or months to reach; but while our forefathers learned to know the country through which they travelled, in all its details, its mountains and valleys, forests and lakes, with us the scenery seems to fly past our vision as we rush ahead with the giant power of steam, and we remember finally hardly more than a few prominent points of what we have seen, or perhaps only a few insignificant details which happened to attract our attention. Likewise, our present civilization reads as it runs and forgets while it is running; a few unimportant details sometimes attract our attention, but the sublimity of the truth is not grasped because we have not yet grown mentally big enough to grasp the whole. Superficial observations are impressed upon our memory and fade again away; only that which reaches the heart, the seat of life, can obtain life. Only that which is known *by the heart* constitutes true knowledge.

We are taught by a superficial science to look upon nature as a compendium of dead things in which, in some incomprehensible manner, life is produced by means of mechanical motion, the origin of the latter being equally incomprehensible. Thus material science teaches that something can be produced out of nothing; *i.e.* that a power can be produced out of a thing in which it does not exist. Judging from her very inadequate and superficial standpoint of observation, she teaches that ~~Life is a product of organization~~, and of the cause of organization she can give no explanation whatever. She believes to know all about "Matter," and yet she knows absolutely nothing about it, except its outward manifestation. *Organization is a product of Life*

We are taught by occult and universal science to look upon the universe as a manifestation of thought, consequently of Consciousness and Life. We are taught by Wisdom, that God and Life, Truth and Power, are *One*, manifesting itself in various forms according to the capacity of the latter, and that man — even the most learned — is himself nothing more than an instrument through which the good or evil powers existing in nature may find their expression.

What becomes of all our scientific and theological self-conceit, of our boasted power and knowledge, if we once fully realize the fact that man is nothing, knows nothing, and has no power of his own, but that all he imagines to have belongs to the universal God, and that he is merely an instrument in which the truth may find its expression or in which it may be misrepresented? No man has a life of his own. The life which he calls his own is merely lent to him during his short appearance upon this earth and must be returned to Nature. It is drawn from the universal storehouse of Life and taken away from him when he disappears from the stage. How little and insignificant is personal man upon this globe, and how great and sublime the majesty of the divine power which manifests itself in Nature, and which may produce its highest manifestation of wisdom and power in Man, if the latter becomes a fit instrument for such manifestations! How small, absurd, and ridiculous appear the disputations of our learned doctors of divinity and philosophy, with all their petty dogmatism and theories, if compared with the supreme Wisdom which the divine element in man may experience if it once becomes self-conscious of its own divinity in the organism of Man! There is only one Truth, and no one can know it except he in whom the truth lives, and who thus becomes an instrument or mirror in which the truth recognizes itself. There is only one true Knowledge and only one *Knower*, and he who desires to obtain true knowledge must become one with the supreme Knower of all. He will then be a magic instrument by means of which the God within knows his own self. The science which deals with the illusions of life is itself illusive, and it is of no value except so long as life's illusions last. It is accessible to the evil disposed as well as to the good, and often those who are the most evil at heart are the most learned and cunning; but the understanding of the fundamental laws of nature, the science of eternal life, is only attainable by a union with the Supreme. The illusions of life may be seen in the light of external nature; but eternal verities can only be seen when the Light of the Logos illuminates the mind.

To know the things which belong to the higher regions of thought, we must be able to rise ourselves to those regions; and as there are few who have the power to do so, the consequence is, that as soon as we begin to speak of such things, misunderstandings arise. Nearly all the theological and philosophical quarrels and disputes among men arise from a misunderstanding of the meaning of terms. So long as men speak of things which can be perceived by the external senses, and which therefore come within everybody's experience, human language is sufficient to enable men to convey their ideas to each other; but when they attempt to build the tower of their ideas into the higher regions of thought, and to speak of things which are beyond the grasp of their intellectual comprehension and beyond their experience, then the Babylonian confusion begins, because each man forms a conception of his own regarding such things, differing from the conception of others; and while all use the same word, yet every one interprets its meaning in a different way. Thus it oftens happens that two persons who are of the same opinion, nevertheless dispute with each other, merely because they differ in their application of terms. Each one denounces the conception of the other as wrong, merely because he himself has formed a wrong conception of what the other believes. It appears therefore, above all, necessary that if we speak of occult or transcendental matters, we should exactly define the meanings of the terms which we desire to use, and we have therefore attempted to give such definitions at the end of this chapter, well knowing that such an attempt will be incomplete, as no amount of logical reasoning can supply the place of the divine power of *Intuition*.

The cultivation of the spiritual power of Intuition is the foremost requirement for the attainment of spiritual knowledge. Intuition is *Reason* pure and unadulterated by any selfish considerations or speculations. It is the power of the *Holy Ghost*, the *Light of Truth*, which shines in the hearts of men. It is a power which is felt in the heart, and if it is cultivated and becomes developed, it grows into a sun which illuminates the whole of the interior man. It has nothing to do with intellectual speculation and logical deduction, which in our modern times are mistaken for Reason; for the Intellect is merely a reflection of the light of Reason; it is like the *Moon* which receives her light from the *Sun*, and which would become dark if the light of the sun were to cease to shine. Man's theories and speculations are his own inventions, but intuition is not invented by man, nor can a man by his own power make himself more intuitional than he is; it is a light from the divine Sun of Grace which descends upon the earth as the rain descends from the clouds.

A bird cannot fly higher than its wings will carry it; a vessel cannot hold more than what it is capable of holding; a man cannot see more than what his organism and means will enable him to see. We have no right to blame the representatives of modern Christianity for not being able to see the true significance of their own symbols, but we advise them not wilfully to shut their eyes to the light of truth when the truth seeks to enter their hearts. We advise them to pay more attention to their own eternal welfare than to the temporal interests of their churches, and to practise the truths which they profess to teach.

The symbols of the ancient Hermetic philosophers have been adopted by the modern Christian church. Many of these symbols have existed from immemorial times, although perhaps the names employed to signify the character of the principles which they represent have been changed many times. The Roman Catholic churches are filled with symbols which were known to the ancient Egyptians, to the still older Brahmins, and perhaps to the inhabitants of the ancient *Atlantis*, whose history fades away in the night of the forgotten past. Priests and laymen bow reverentially before those signs; because they are taught to do so and because they consider them as historical mementos of the death of Jesus of Nazareth.

The modern skeptic treats those signs with contempt. How could he reasonably do otherwise? for the conceptions which he has formed about them in his own mind are absurd and childish, and moreover they remind him of the days of "religious" persecution and intolerance. Little does the Christian or the skeptic know about the true significance of these symbols; for their real meaning has never been taught them. We do not know of a single book issued by any modern Christian authority giving their true explanation. Their secret meaning was known to the ancient Initiates into the Mysteries; but the modern keepers of these symbols have lost the key to their understanding and know little more about them than their external forms.

Little indeed would be the value of these figures, if they had no other meaning but that which is represented upon their surface. Little indeed would be the value of the Bible, if its stories referred to nothing else but to events said to have taken place in the history of the Jews. Even if such events had actually taken place as described, they could be of little interest to us, because that history belongs to the past, and the persons concerned in it are gone from the stage of the world. But if we look with the eye of the spirit within the unseen world and learn to observe the processes going on within the realm of the soul of the Universe, whose true image and representation exists within our own selves, we soon learn that the fables and allegories represented by such tales and symbols are of a deep significance and their knowledge of supreme importance to us; and moreover it seems that many of these stories have been made purposely so absurd that no reasonable person should be tempted to accept them in their literal sense.

As the images of the gods of the Greeks and Romans were not intended to represent persons, but consisted merely of figurative personifications of universal powers of nature; likewise the persons spoken of in the Bible are personifications of the same powers or principles, which still exist to-day as they existed before the Bible was written. We meet them again in the sacred book of the East, in the *Mahabharata* and *Bhagavat-Gita*, in the songs of *Homer*, and in other inspired books. They are mythological allegories, but they are therefore not less true, because, if properly understood, they represent living truths.

The surest sign of the decay of a religion is when the secret meaning of its symbols becomes entirely lost. It is a sign that the spirit which gave life to that religion has fled, that opinion has taken the place of *Faith*, and belief the place of Knowledge. The external form of such a system may exist for a while, but finally the dead form will dissolve in spite of all efforts to prevent its decay.

The universal belief in the external significance of the symbols of the ancient Egyptians, Romans, and Greeks has been a precursor of the decay of those religions; the continued disregard of the true meaning of the symbols of the Christian churches will surely lead to the decay and dissolution of the latter. This decay is so universally visible and publicly complained of and acknowledged by the professors of the Christian religion, that it would be a waste of words to attempt to prove that which no one denies.

What greater service could therefore be rendered true religion than to restore their true meaning to the sacred symbols of the past, and to induce those who desire the truth to study the signs by which the fundamental laws of physical and spiritual evolution have been represented far better than could possibly be done by a verbal description? Words are misleading; they are useless to him who has no intuition; but to him who possesses the power of thought, a point, a line, a triangle, or a circle is very suggestive, and may be sufficient to indicate to him the way to arrive at the truth.

Will the publication of the *Secret Symbols of the Rosicrucians* help to restore the crutches to a decrepid church, or will those who are blind reject the truth, if it does not come to them under the seal of some man-made authority? Will this book serve to infuse life into a corpse, or will the corpse be permitted to putrefy and a new child be born? Will the new wine be poured into new bottles instead of being filled into old casks whose staves are crumbling to pieces? Will priestcraft grow again into power upon this earth; or is the time approaching when every man will be a king and a priest in his own realm, a conqueror over self, free from sectarian and scientific prejudices, worshipping no other god but the Truth, and professing no other creed but the love of divine Reason and the Unity of the All?

MYSTERIUM MAGNUM STUDIUM UNIVERSALI.

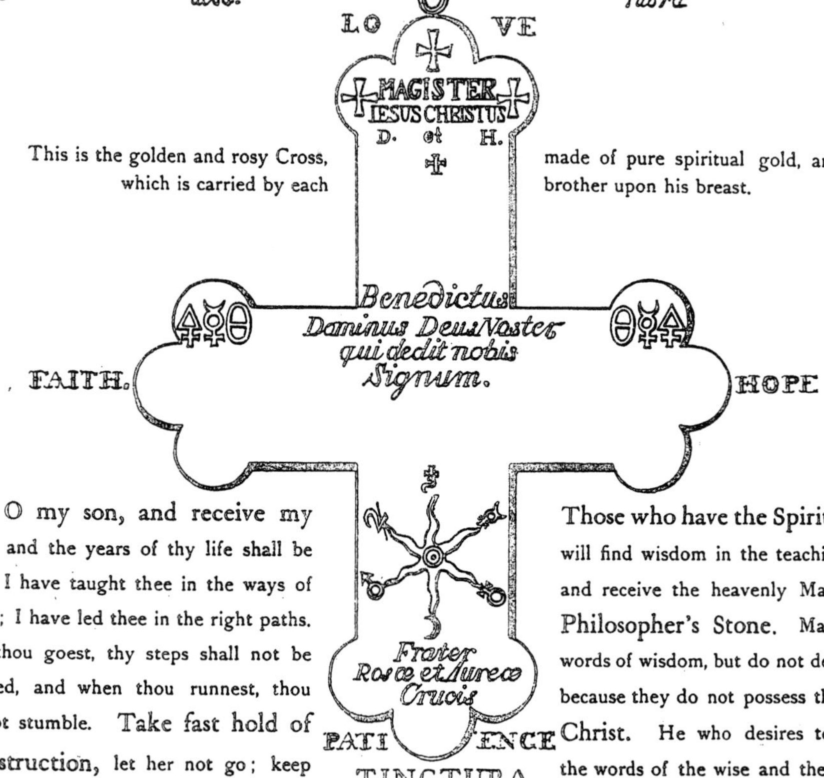

This is the golden and rosy Cross, which is carried by each made of pure spiritual gold, and brother upon his breast.

FAITH. HOPE

Hear, O my son, and receive my sayings, and the years of thy life shall be many. I have taught thee in the ways of wisdom; I have led thee in the right paths. When thou goest, thy steps shall not be straitened, and when thou runnest, thou shalt not stumble. Take fast hold of the instruction, let her not go; keep her, for *she is thy life.—Prov. IV. 10.*

Those who have the Spirit of Christ, will find wisdom in the teachings of Christ and receive the heavenly Manna and the Philosopher's Stone. Many hear the words of wisdom, but do not desire wisdom, because they do not possess the Spirit of Christ. He who desires to understand the words of the wise and the doctrines of Christ, must become Christ-like himself.

"Call unto me, and I will answer thee, and shew thee great and mighty things."—*Jeremiah xxxiii.*

PART I.

ELOHIM
JEHOVAH
GOD
WORD
Fiat
Natura
Primum Mobile.
Prima Materia.
Quinta Essentia.
Quatuor Elementa.
Lapis Philosophorum.
FIRE.

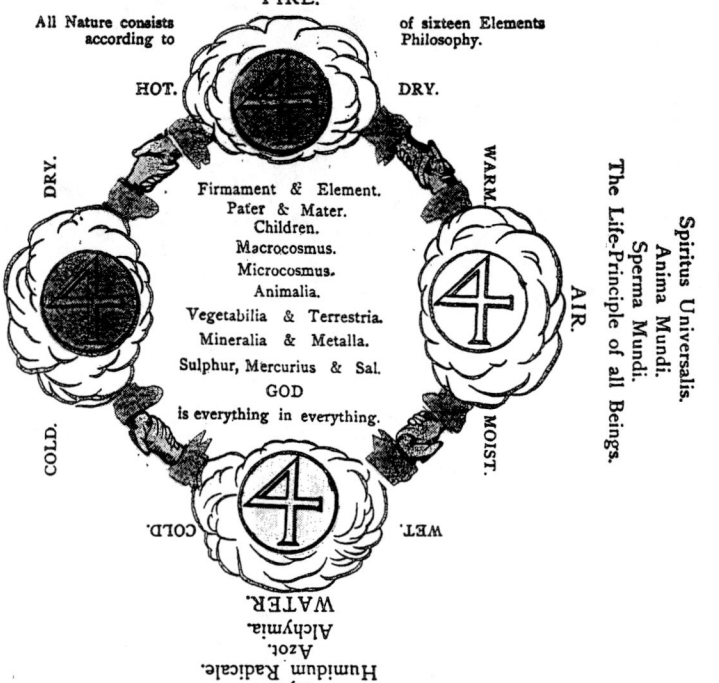

All Nature consists according to — of sixteen Elements Philosophy.

HOT. DRY.

DRY. WARM.

Firmament & Element.
Pater & Mater.
Children.
Macrocosmus.
Microcosmus.
Animalia.
Vegetabilia & Terrestria.
Mineralia & Metalla.
Sulphur, Mercurius & Sal.
GOD
is everything in everything.

EARTH. AIR. MOIST. CHAOS.

Rebis. Spiritus Universalis.
Sal. Anima Mundi.
Chimia. Sperma Mundi.
Corpus. The Life-Principle of all Beings.

COLD.

WET. COLD.
WATER.
Alchymia.
Azot.
Humidum Radicale.
Hyle.

"For this they willingly are ignorant of, that by the Word of God the heavens were of old and the earth standing out of the water and in the water." *II Pet. iii. 5.*

"The earth was without form and void and darkness was upon the face of the deep, and the Spirit of God moved upon the face of the waters." *I Genes. i. 2.*

Ignis Philosophorum.	Aqua Philosophorum.
invisibilis and secretissimus occultatum	Mercurius Primaterialis Catholicus.

Desire the Fire,
Seek for the Fire:
And you will find the Fire,
Kindle a Fire,
Add Fire to Fire,
Boil the Fire in the Fire,
Throw Body, Soul and Spirit into the Fire:
And dead or alive you will possess the Fire,
This Fire becomes a black, yellow, white and red Fire,
Give birth to your children while you are in the Fire,
Feed, water and nourish them in the Fire:
They will then live and die in the Fire,
They will be Fire and remain in the Fire,
Their Silver and Gold turns to Fire.
Heaven and earth will perish in Fire
And at last there will be a four-fold *Philosophical Fire*,
 i. e., the *Celestial Fire*.

4 × 4 = 16 lines,
and this is the number of the
ELEMENTS.

Water is Water and always remains Water;
From the heavens of the wise comes Water;
The Philosopher-stone weeps Water,
But the world does not know or esteem that Water.
The Fire of the wise burns in Water
And lives in the Water.
Turn the Fire into Water
Boil the Fire in the Water:
And you will obtain a fiery Water,
An acrid Water, resembling Sea-Water.
To the children it is a living Water,
It turns body and soul to Water.
Becomes fetid, green, foul, blue, like heavenly Water.
Digest, calcinate, dissolve and *putrefy* this Water.
Seek the four-fold, imperishable, *Philosophical* Water.
And when you have succeeded, your art will turn to water,
 i. e., the *Mysterious Water*.

4 × 4 = 16 lines,
and this is the number of the
ELEMENTS.

The Mystery of △ the Number

THREE.

"Stand in awe and sin not." "Commune with your own heart upon your bed and be still."—*Psalm iv., 4.*

Try to find out the spiritual significance of the triangle, and learn to know thyself.

The Divine Light   becomes manifest in the   Philosophical Light
of Grace   of Nature.

The Spiritual  Sun   

The Root Jesse,   The Root of all Metals.

Study the relations of the numbers Three and Seven, and the manifold allegories connected with these numbers in the Bible.

"And the Lord said unto Abram: Take me an heifer of three years old ⎫
                       and a she goat of three years old ⎬ The Three Principles.
                       and a ram of three years old, ⎭
                       and a turtledove and a young
                           pigeon.—*The Gluten of the White Eagle.*

And he took unto him all these, and divided them in the
    midst, and laid each piece one against another, - - *Solutio Philosophorum.*
but the birds divided he not; - - - - - - - - *Sophist Separatio.*
and when the fowls came down upon the carcases, Abram
    drove them away." - - - - - - - - - - - *Caput Mortuum.*

But the bird of Hermes ☿ eats the dead carcases and flies away. It is caught by the Philosopher, strangled and killed.

Try to find spiritually the Fire and the Water, which is the **Prima Materia** or the **Spiritus Universi**, in which the Gold is consumed and from whence the latter, after the **Putrefaction** is over, resurrects into a new life.

Try to find out what are the three **Births**, and what are the three principles of man;

Spirit, Soul, and Body,

In the Light of Divinity and in the Light of Nature.

The following figure is one of the Keys:

The Mystery of the Number
FOUR.

Happiness.
The Light of Nature.

Light of Grace.

Woe! Woe! Woe!
to the Sophists.

Quinta Essentia.

4 Elements.	3 Beginnings	2 Seeds.	1 Fruit.
4 Fire △ 1	Sulphur 1	Male ☉	Natural product 1
3 Air △ 2	Salt ⊖ 2	Sperma 2 Sem. 2	Tinctura
2 Water ▽ 3			
1 Earth ▽ 4	Mercury 3	Female ☽	Supernatural 2
of God	of Nature	of Metals	of Art
Father	Son	Holy Ghost	Christ & Man
G.	P.	W.	M.

He who understands this figure, will see how one thing originates from another one. Everything exists fundamentally of four elements. They produce three Beginnings, and from these originate two sexes, Sun and Moon, but the latter two produce the Son, the mortal and the divine Man

Smoke will arise above your heads from eternity to eternity, and be a torture to you.

Try to find out spiritually the secret signification of the Number Four, which is alluded to frequently in the allegories of the Old and New Testament.

The number 40 is frequently mentioned in the Bible.

The Key to the explanation is:

3 x 4 x 40.

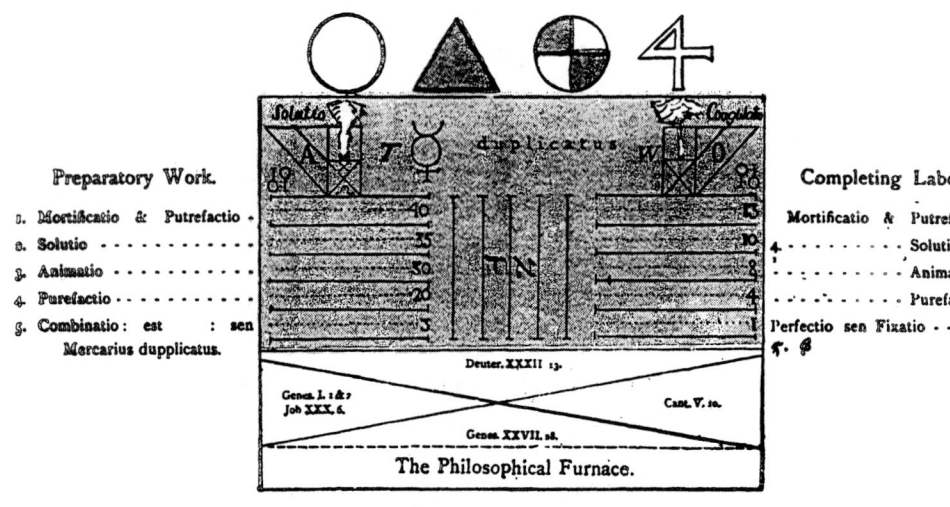

Preparatory Work.

1. Mortificatio & Putrefactio
2. Solutio
3. Animatio
4. Purefactio
5. Combinatio : est : sen Mercarius dupplicatus.

Completing Labour.

Mortificatio & Putrefactio 1.
Solutio 2.
Animatio 3.
Purefactio 4.
Perfectio sen Fixatio 5.

The Philosophical Furnace.

The Mysteries of the Number SEVEN.

In this mystery is the ness, Peace and Tran Eternity. It is the erated Children of

THE ROSY

In its theosophical and

The Secret Cross, which although the world talks

Eternal Life, Happiquility in Time and in Heaven of the regen- God.

CROSS

theological aspect;

is unknown to the world, much about it.

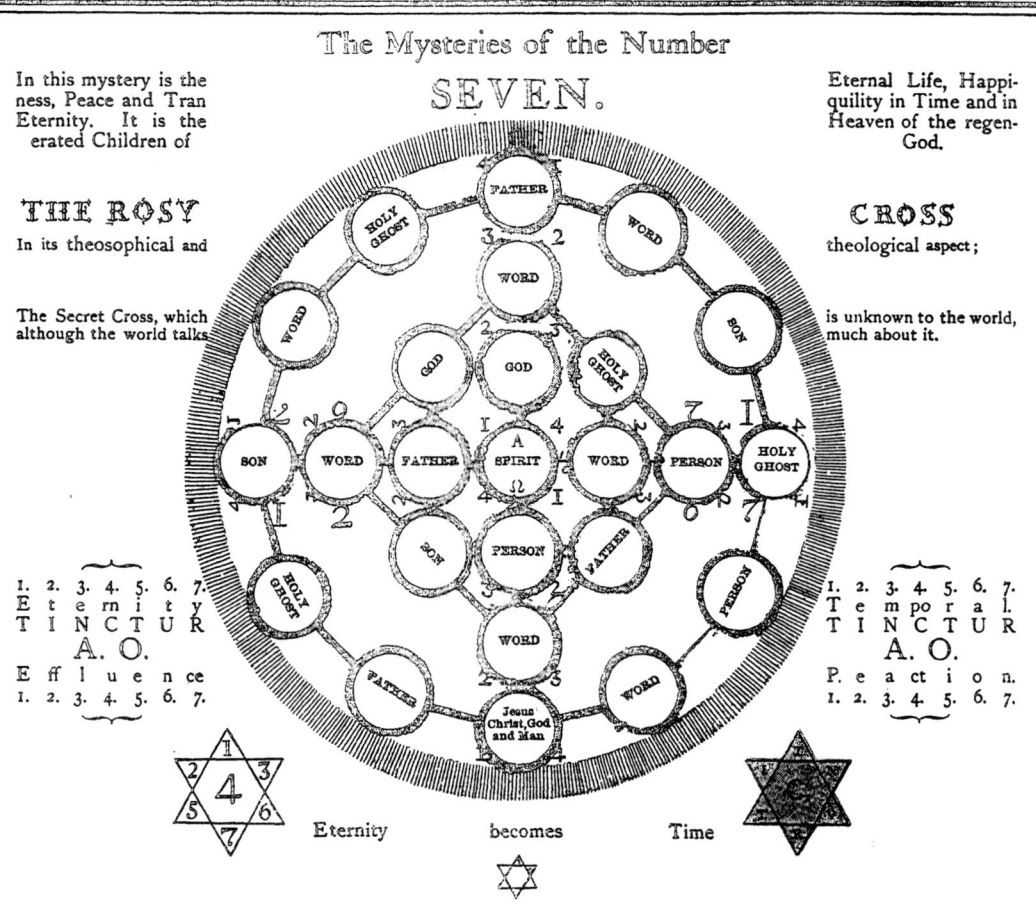

I. 2. 3. 4. 5. 6. 7.
E t e r n i t y
T I N C T U R
A. O.
E f f l u e n c e
I. 2. 3. 4. 5. 6. 7.

I. 2. 3. 4. 5. 6. 7.
T e m p o r a l
T I N C T U R
A. O.
P. e a c t i o n.
I. 2. 3. 4. 5. 6. 7.

Eternity becomes Time

In this symbol is represented Eternity and Time, God and Man, Angel and Devil, Heaven and Hell, the old and the new Jerusalem, with all beings and creatures, time and hours contained therein

12 Patriarcha. 12 Stars in the Crown.
12 Propheta. 12 Heavenly Signs.
12 Apostles. 12 Months in the Year.
12 Articles of their Faith. 12 Hours of Day.
12 Gates of the New Jerusalem. 12 Hours of Night.

Probat Fidem.

There is a Word speaking eternally. It speaks itself from within but not in itself, and yet it can never be spoken. This Word is I, All, Everything and Nothing; Heaven and Hell, Light and Darkness, Good and Evil, Spirit and Matter, Will and Desire, Joy and Sadness, the Real and the Illusive, Time and Eternity, Angel and Devil, Life and Death, Sound and Stillness, Something and Nothing, Man and God ; All in All ; called the Christ.

It cannot approach itself, and it cannot be compared to anything. It speaks, speaks not and is spoken, speaks out and in and remains unspoken. It creates everything, remains forever uncreated and is itself everything which it created.

It was, is, is not and yet will be One God, one Lord, one Spirit, one Unity. Acting from within outwardly, and from the Outward to within.

He who cannot believe this will not comprehend the rest. From Faith grows knowledge and knowledge produces Faith. First believe, then TRY. If you find it good, praise it.

Silentium Sapientiæ ; Simplicitas Veritatis.

SIGILLUM.

The Cross is the best exposition of the Holy Scripture.

CONSTANTIA.

Animæ Pharmaca
Sanctissima Bibliotheca
Lecta placent. Xies repetita placebunt
Via Sancta
SPIRITUS & VITA
Oraculum & Spiraculum
Jɛ Ho Væ
Rationale Divinarum
A.O U R I M & T U M M I M
Tabernaculum
DEI cum Hominibus
S A N C T U A R I U M
יְהוָה
MEMORIALE
Magnalium DEI
LUCERNA DOMINI
Armarium
Spiritus Sancti
P A N A C E A
Nectar & Ambrosia
PORTA COELI
LIBER DOMINI
FONS
Signatur
CIBUS ANIMÆ
Lumen Gratiæ
ORTUS
Conclusus
☩ ESAURUS
Absconditus
VERBUM VITA
Querite   Invenietis
Credite & Intelligetis.

The Great Mystery,

which has been hidden from the world, but which has become revealed by God in His Saints, to whom it is known that the great and inestimable treasure to a Christian (meaning Christ in Man. Col. i., 27) is the true spiritual perception and recognition of Jesus Christ; in other words, of God in Man and of all celestial and terrestrial wisdom in Heaven and upon the Earth.

G.P.W. F.S.H.G.
G.M.
J. C.
Eternity becomes Nature and Time.
Materia Prima becomes
Materia Ultima.
This is life eternal, that they might know thee the only true God and Jesus Christ whom thou hast sent. John XVII.
He who has Christ (Man and God) in him, has nothing to fear.
Cling to the divine Man, the eternal Christ whose nature is dual. God and Man.
Divine Cabalistic Signet Star.
The seven Divine Spirits.
Natural Philosophical Stone.
The seven Metal Spirits.
FATHER
Son. Spirit.
One God, a Unity of whose Light was born the Christ.
1. 2. 3 4
God
Eternity
Sulphur, Tria
Mercurius, Sal, Principia.
I H E S U S
A D O N A I
J E H O V A
K R E S T O S
T I N C T U R
1. 2. 3. 4. 5. 6. 7.
C H A O S
Grace be with all them that love our Lord Jesus Christ in sincerity Ephes. VI. 24.
I determined not to know anything among you, save Jesus Christ and him crucified. John II. Corinth. II.
Eternal God
Physical Nature.
who was in the beginning, the Word: which we have seen with our eyes.
which we have seen and touched with our hands John I
God dwells in the Light, which no Man can approach, which no Man can approach unto. I Tim. VI. 16
God is Spirit and must be worshipped in Spirit. John IV. 24
No one knows the Son but the Father and no one knows the Father but the Son, and he to whom the Son will reveal him.
SON
With the Father & H. Ghost a Unity and God, is called Man because the Word has become manifest in the flesh A Trinity in eternity and in time II John 10.
God was in Christ and reconciled the world with himself. God was killed after the flesh, but made living after the Spirit.
GOD
The Eternal Spirit, the cause and source of all things, radiating his Light (Son) into the world and yet remains God in Heaven and upon the Earth.
No one enters Heaven except the Son of Man who came from Heaven. That which is born of the Spirit is Spirit, that which is born of flesh is flesh.
H. GHOST
is Father and Son Emanates from the Father and the Son, and yet is only one Spirit, one Father, one Christ and son of Man in time and eternity.
The Son of God becomes manifest in the flesh.
VERBUM incarnatum.
Anima Mundi Spiritus Coelest.
RAD. HUM. Tinctur.
Id quod fuit permansit, quod non erat assumsit. In Christ dwelleth all the fulness of the Godhead bodily. Col. II. 9
To the Christ in Man is given all the power in Heaven and upon the Earth.
In the flesh resides corporeally the plenitude of Divinity.
LUX
FIAT CORPUS.
C A O S
1. 2. 3. 4.
2 Corp. 2 Spirit.
Water turned to Stone.
N.B. The Life has appeared, we have seen it and are witnesses of it, and we know that it is eternal. The love of Christ is better than all (superficial) science.
N.B. Nec non primarum Materiarum, i.e. Principium omnium rerum, five Tinctura Lapidis Philosophorum. If any man love not the Lord Jesus Christ, let him be anathama maran-atha. II Cor. XVI. 22.
A.O.I.C.
God and Man.
The eternal Light radiating into the world becomes flesh and feeds the flesh with its own substance transforming it into a new being.
MAN
The One
the Word
Flesh.
TIME
Triune God
becomes
A
God and Man
H E
E T
(Eternity) (Time)
V
O
Heaven.
Light.
Darkness.
C R I T U S
Hell.
The Cross is the greatest miracle It exists in God and in Nature.
I.G G C.
Divine Cabala Quinta Essentia.
W P
M.
Kill the old Adam and his evil desires.
Destroy the 1. 2. 3. 4. Elementa and their evil emanations.

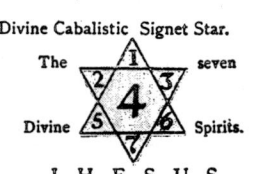

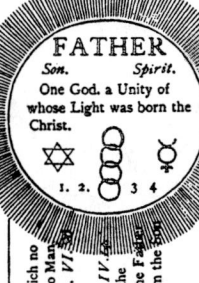

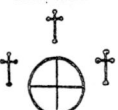

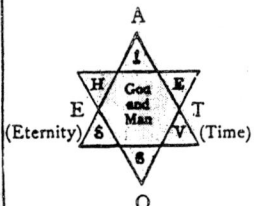

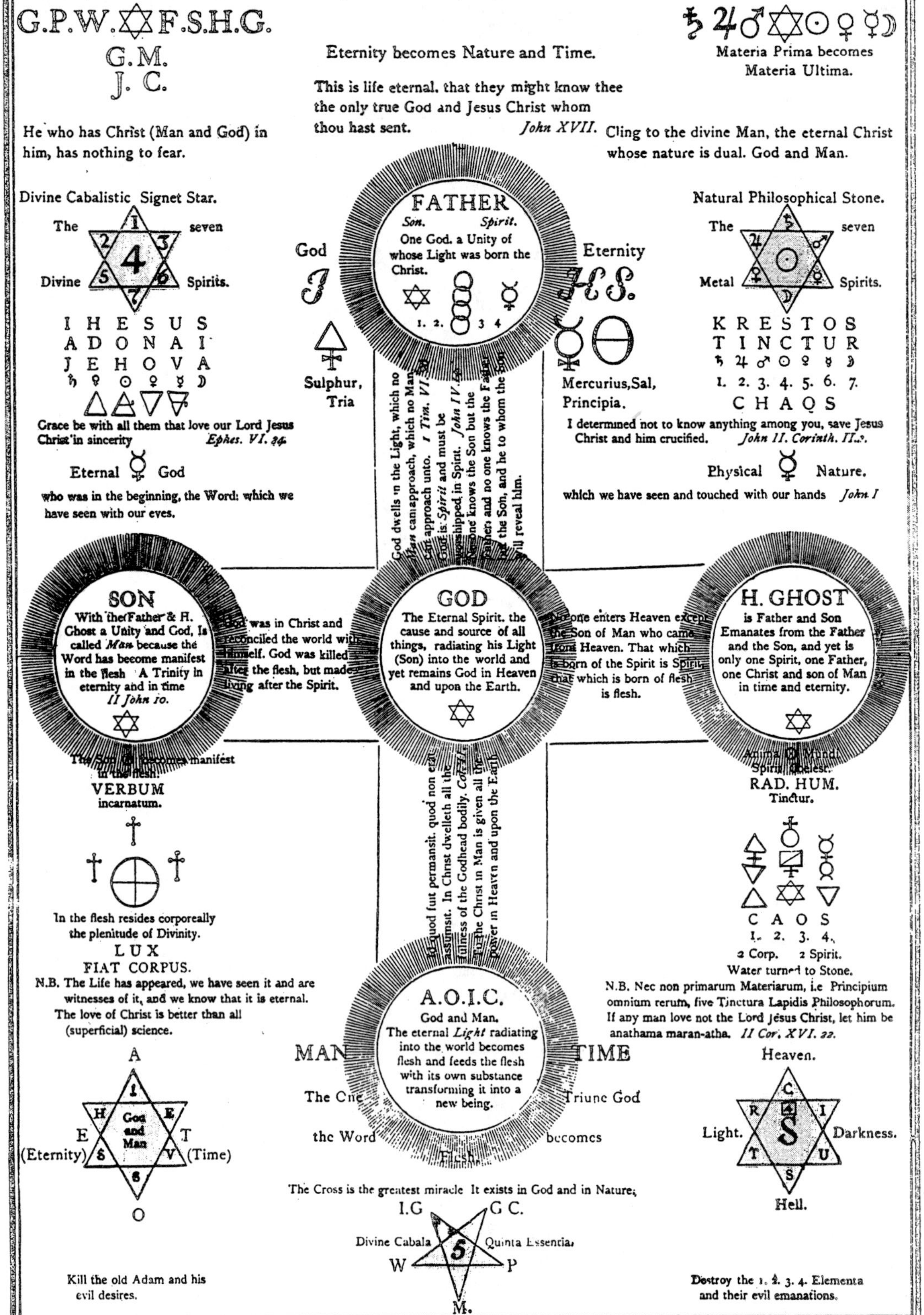

FIGURA CABALISTICA.

The Basis of the wonderful numbers:

1. 2. 3. 4.
ELOHIM.

Arcana Arcanorum
The One Eternal God becomes manifest in Trinity

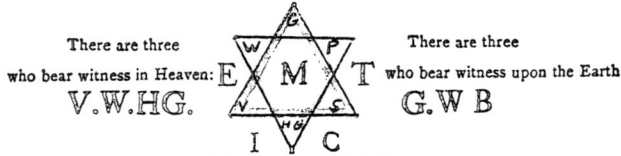

There are three who bear witness in Heaven: V.W.HG.

There are three who bear witness upon the Earth: G.W.B

and the three are only one;

In their eternal aspect **heavenly**, in their terrestrial aspect **natural**.

This is In Heaven and upon the Earth.

The complete 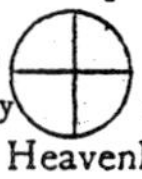 Rose-Cross visibly manifest and yet the greatest mystery of all mysteries in heaven and on earth.

Eternally Divine.	Heavenly.	Natural and finite.
In the eternal divine Light, A. Ω. Theosophia. Mysterium Magnum according to the divine Cabala.	In the heavenly Light and Corner-stone G. and M. J. C. THEOLOGIA.	In the natural Light and the Philosopher's Stone Philosophia. Mysterium Magnum according to the Magia Philosophia.

The seven heavenly principles and their qualities. — Tinctur. An eternal omnipotent God.

The seven terrestrial principles and their qualities. — Tinctur. A small terrestrial but powerful God.

Explanation according to the A and O.

One God {
  1. Spirit, 2. Person, 3. Word, — 3 eternal spiritual celestial persons — in one being
  1. Father, 2. Son, 3. Holy Ghost, — 3 celestial persons manifest in time — in one being
  1. God, 2. Christ, 3. Man, — 3 celestial and 3 terrestrial persons — in J. C. the perfect Man.
}

The one, only and triune God is a prototype of the Universe, with all that is contained therein, in

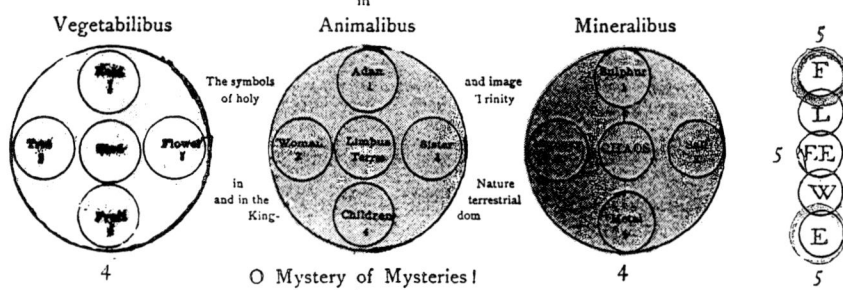

Vegetabilibus — Animalibus — Mineralibus

The symbols of holy and image Trinity in and in the King- Nature terrestrial dom

O Mystery of Mysteries!
He who truly knows Christ has well employed his time.

Note.—The secret meaning of the numbers 1, 2, 3, 4, explains the true Rose ⊕ Cross, the divine manifestation of God in Man. In it is contained all celestial and terrestrial Wisdom in Heaven and upon the Earth. God is generated and born out of itself a Trinity and yet a Unity, spiritual, celestial, invisible, in Eternity, (1. Spirit or God; 2. Word; 3. Father) and corporeal, terrestrial and visible in Time, a Duality of Man and God (1 Spirit; 2. Person; 3 Word). The Word became Flesh; Eternity became Time; God became Man; that is to say: One time, two times and half a time, according to the old and the new Testament; the celestial and terrestrial Trinity in its plenitude in Heaven and upon the Earth (*)

(*) See Franz Hartmann "In the Pronaos of the Temple of the Rosicrucians" and Cornelius Agrippa "Philosophia Occulta."

PART I.

The Philosopher's Stone.

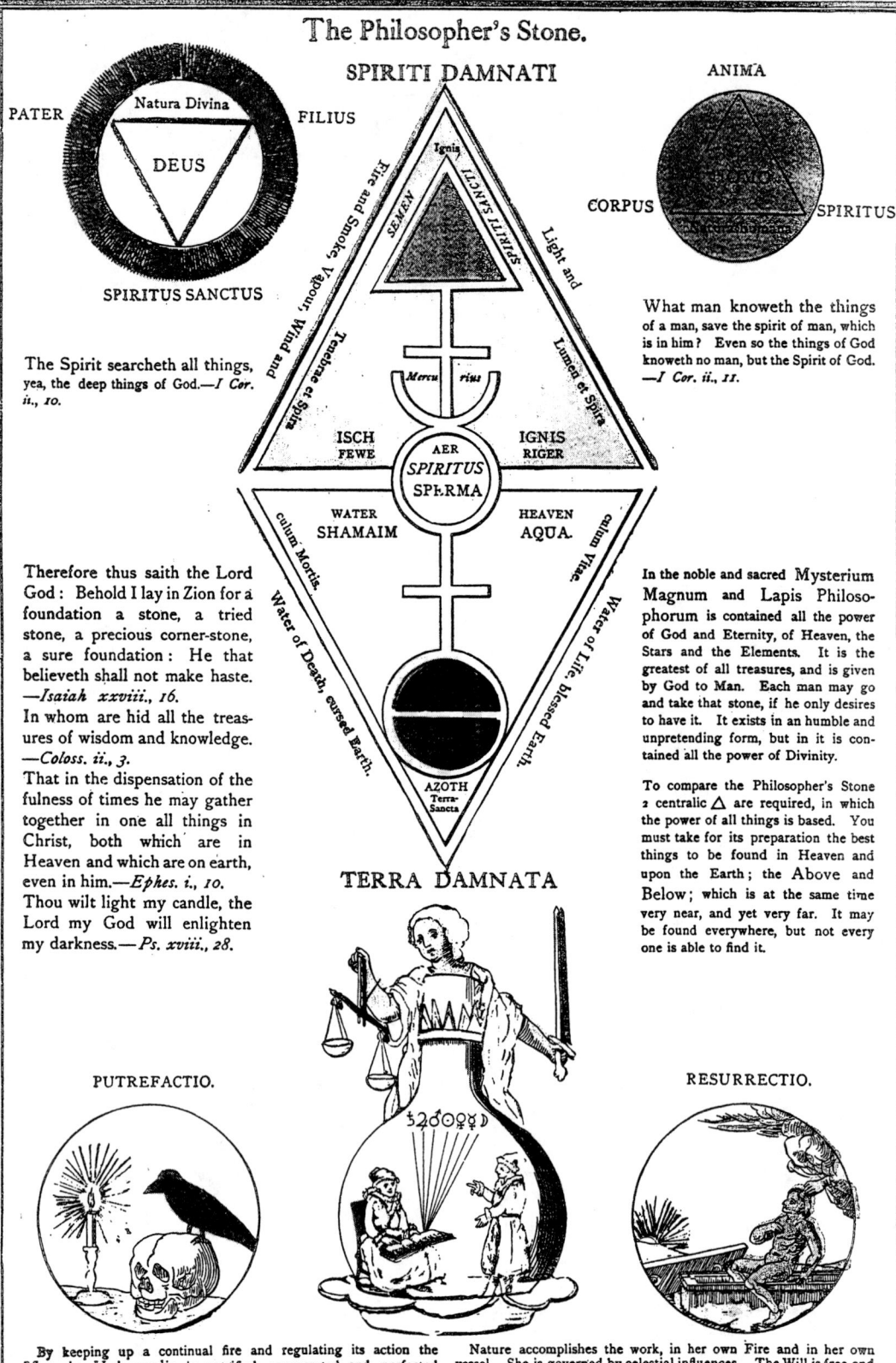

The Spirit searcheth all things, yea, the deep things of God.—*I Cor. ii., 10.*

Therefore thus saith the Lord God: Behold I lay in Zion for a foundation a stone, a tried stone, a precious corner-stone, a sure foundation: He that believeth shall not make haste.—*Isaiah xxviii., 16.*

In whom are hid all the treasures of wisdom and knowledge.—*Coloss. ii., 3.*

That in the dispensation of the fulness of times he may gather together in one all things in Christ, both which are in Heaven and which are on earth, even in him.—*Ephes. i., 10.*

Thou wilt light my candle, the Lord my God will enlighten my darkness.—*Ps. xviii., 28.*

What man knoweth the things of a man, save the spirit of man, which is in him? Even so the things of God knoweth no man, but the Spirit of God.—*I Cor. ii., 11.*

In the noble and sacred Mysterium Magnum and Lapis Philosophorum is contained all the power of God and Eternity, of Heaven, the Stars and the Elements. It is the greatest of all treasures, and is given by God to Man. Each man may go and take that stone, if he only desires to have it. It exists in an humble and unpretending form, but in it is contained all the power of Divinity.

To compare the Philosopher's Stone 2 centralic △ are required, in which the power of all things is based. You must take for its preparation the best things to be found in Heaven and upon the Earth; the Above and Below; which is at the same time very near, and yet very far. It may be found everywhere, but not every one is able to find it.

By keeping up a continual fire and regulating its action the Materia Universalis is putrified, regenerated and perfected within one vessel and furnace. Nature herself does the work by means of her own inward Fire; but the latter is stimulated by the Philosophical Fire. No other work is required from the Alchemist, to perform, but to keep up the Fire by Aspiration and Prayer.

Nature accomplishes the work, in her own Fire and in her own vessel. She is governed by celestial influences. The Will is free and may towards the end of the work put certain limits to Nature, so that she may not go too far. The Will, being free, governs Nature, by attracting her, but if the Will is attracted by Nature, it ceases to be free. Nature will then counteract the Will and destroy the work.

PART I.

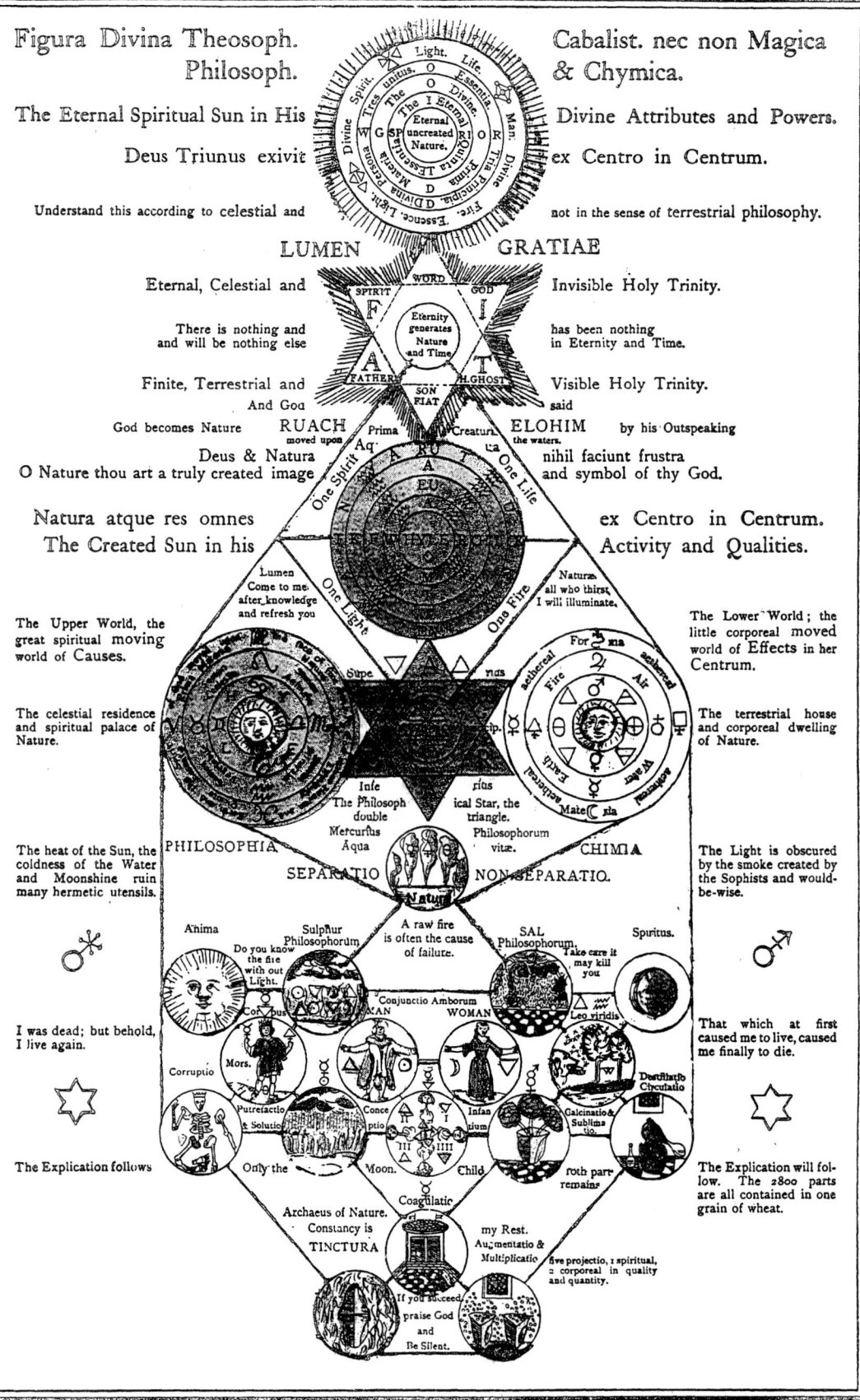

Figura Divina Theosoph. Philosoph. — Cabalist. nec non Magica & Chymica.

The Eternal Spiritual Sun in His Divine Attributes and Powers.

Deus Triunus exivit ex Centro in Centrum.

Understand this according to celestial and not in the sense of terrestrial philosophy.

LUMEN GRATIAE

Eternal, Celestial and Invisible Holy Trinity.

There is nothing and has been nothing and will be nothing else in Eternity and Time.

Finite, Terrestrial and Visible Holy Trinity.

And God said

God becomes Nature RUACH moved upon Prima Creatura ELOHIM the waters. by his Outspeaking

Deus & Natura nihil faciunt frustra

O Nature thou art a truly created image and symbol of thy God.

Natura atque res omnes ex Centro in Centrum.
The Created Sun in his Activity and Qualities.

The Upper World, the great spiritual moving world of Causes. — The Lower World; the little corporeal moved world of Effects in her Centrum.

The celestial residence and spiritual palace of Nature. — The terrestrial house and corporeal dwelling of Nature.

The heat of the Sun, the coldness of the Water and Moonshine ruin many hermetic utensils. — The Light is obscured by the smoke created by the Sophists and would-be-wise.

PHILOSOPHIA — CHIMIA
SEPARATIO — NON-SEPARATIO.

A raw fire is often the cause of failure.

I was dead; but behold, I live again. — That which at first caused me to live, caused me finally to die.

The Explication follows — The Explication will follow. The 2800 parts are all contained in one grain of wheat.

five projectio, 1 spiritual, 2 corporeal in quality and quantity.

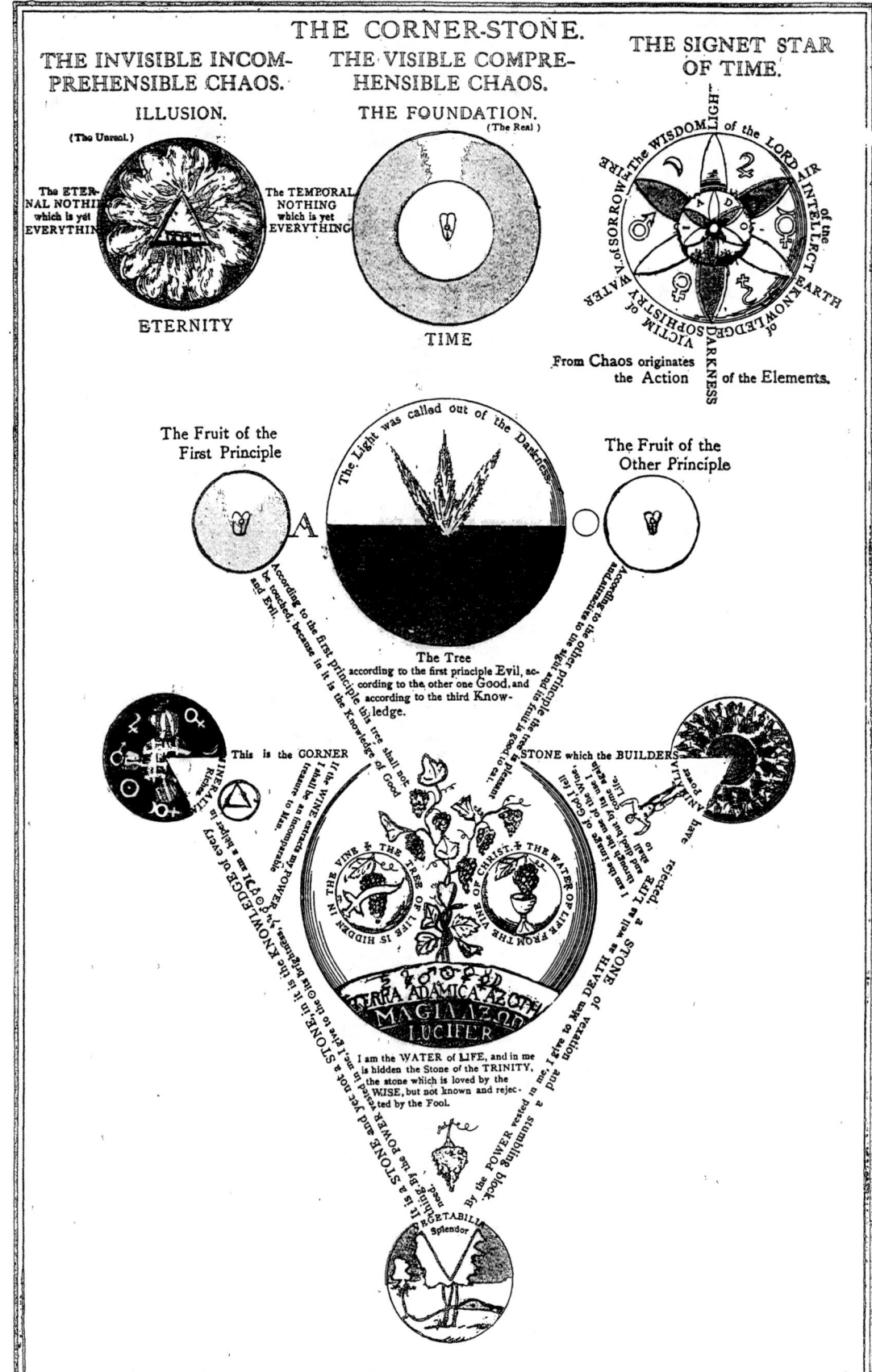

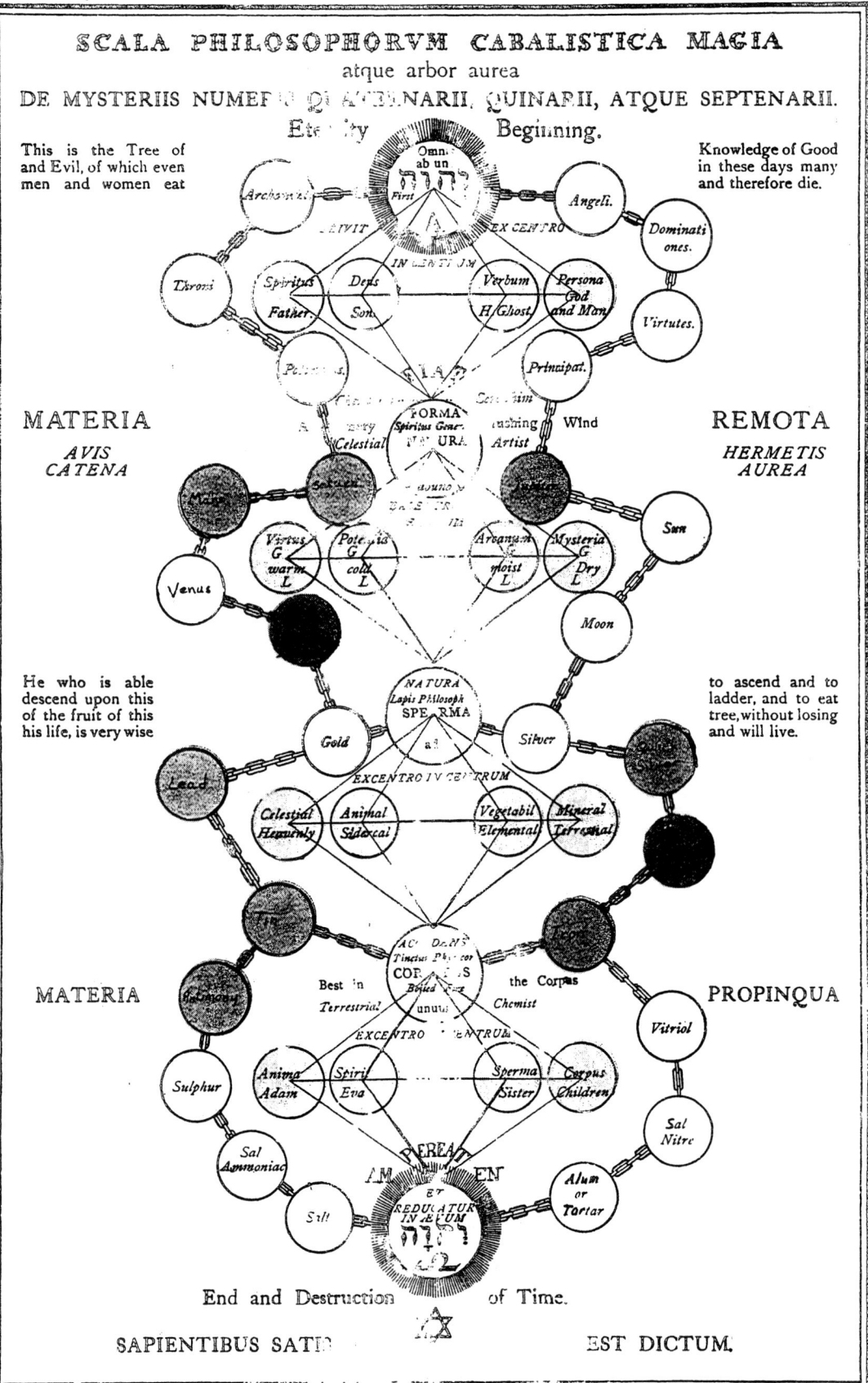

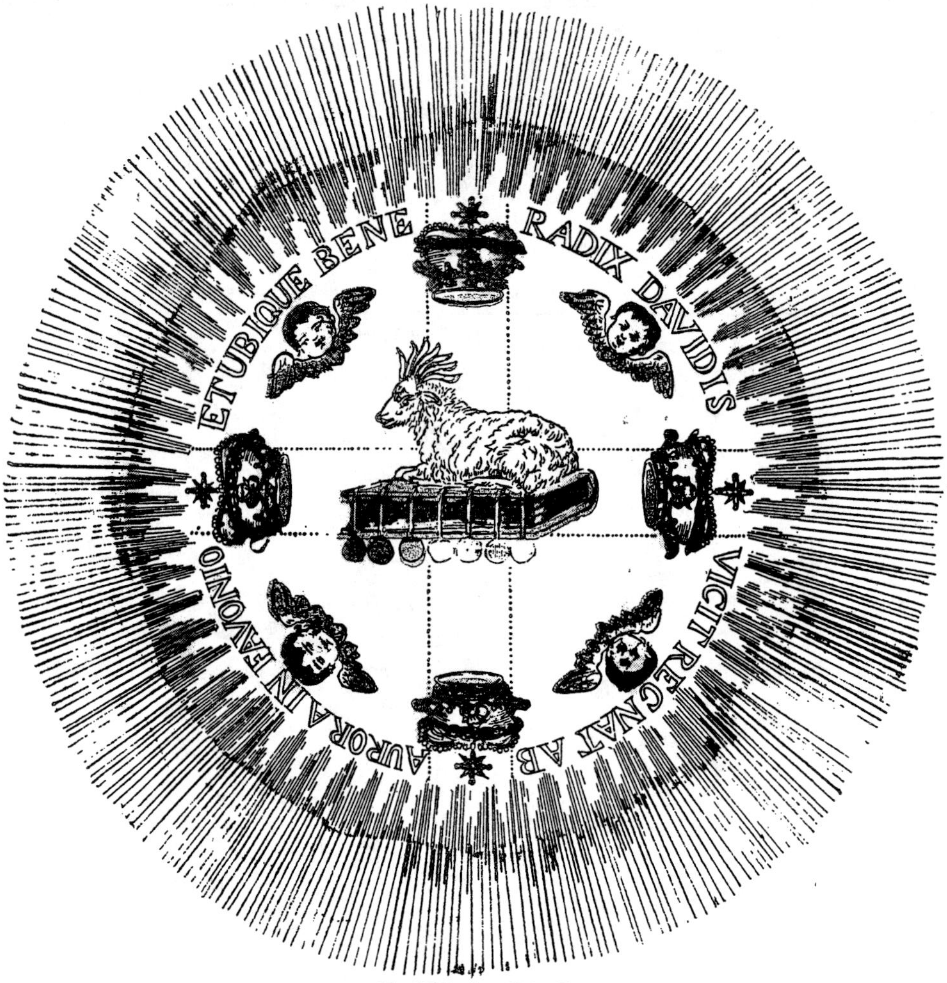

JESUS.

I desire no other Knowledge, no other Power, no other Love, I have no other Joy or Ambition or Desire, neither in Heaven nor upon the Earth, except that which comes from the living WORD, which has become Flesh in Man.

This is the most wise and holy article of faith, having been revealed by
GOD in the LIGHT of NATURE.

Physica.   I Am the A and Ω   Metaphysica
          the First and the Last.   and Hyperphysica.
Apocal. I. ii. 12. Cap. V. 5. seq.

D. O. M. A.

Deo omnipotenti fit Laus, Honos & Gloria in Seculorum Secula, Amen.

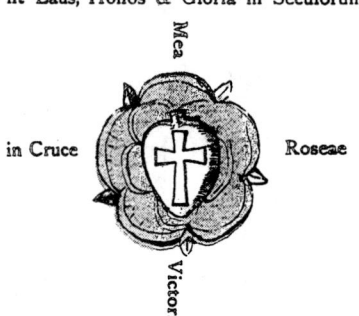

A Sermon

rendered by an unknown Philosopher, a Brother of the Fraternity (R. C.) and containing the Substance of the teachings of the Sacred Philosophy and the process of preparing the Universal Panacea.

THE TRIUNE GOD, called by men JEHOVAH

produced every Thing out of No Thing
The Spirit of God moved upon the face of the Waters, the CHAOS,
which is the primum HYLE of the Sages, or the Substance (Water) out of which everything has been formed,
Firmament, Mineralia, Vegetabilia, Animalia: called

THE MACROCOSM,

and from its Centre and Quint-Essence

THE MICROCOSM,

being the most perfect being, produced by the producer of all things, and called

MAN,

male and female in one, a spiritual power, an image of God:
An immortal Soul, a heavenly invisible Fire.

He fell into Matter and Darkness, but there is the MESSIAH, (which means:)
The Spiritual Light of Grace in the Light of Nature.
LILI: The Prima Materia of a perfect Body,
the matrix or womb of the Middle World,
Balsam and Mumia,
and the incomparable magical Magnet in the Microcosm,
The WATER of the sages, from which all things come, and in which all things are contained,
and in which imperfect things are improved
A Healthy Soul in a Healthy Body,
Indefatigable prayer, patience and perseverance are required
The Substance, the Alembic, the Furnace, the Fire are all only one thing,
contained in only one, and this One is itself the Beginning, the Middle and the End.
It suffers nothing foreign and impure to exist in it; it is prepared without any admixture of foreign matter;
for see: In the MERCURY is contained all that the Wise desire:

THE CLEAR AND TRANSPARENT FOUNTAIN.

The Double Mercury.
A revolution of the sphere of all the planets
and a Being, which in a moment of time emits a black smoke of a

LUMINOUS
DEATH AND LIFE.

Regeneration and Renovation,
being the Beginning, the Middle and End of Finity or Permanency,
the noblest of all Secrets, the Foundation of the whole magic mystery
Take the Quint-essence of the Macrocosmos and Microcosmos, the Philosophical MERCURY,
the invisible, celestial and living Fire
and the Salt of the Metals;
prepare of this according to the rules of Magic, by Rotation, Solution, Coagulation and Fination

THE SUPREME UNIVERSAL PANACEA

wherein exists
Supreme Wisdom, perfect Health, sufficient Riches.
All comes from ONE, and All becomes ONE
in the end
Reject all evil desires, because they are an impediment in this work.
Let the will of JEHOVAH be done,
it will accomplish everything

To God alone belongs all Honor and Praise.

Lege.

Judica.

Tace.

VOCABULARY OF OCCULT TERMS.

WRITTEN FOR THE PURPOSE OF MITIGATING THE CONFUSION CREATED BY THE BUILDING OF THE TOWER OF BABYLON.

"*Omnia ab Uno*" is one of the mottoes of the Rosicrucians. It expresses the idea that the All has been evolved from One; or, in other words, that God is one and indivisible, and that the multifarious activities of life which we see in the universe are merely various forms of manifestations of God; or, to express it more correctly, of the *creative Power*, the *Light* and *substance of Life*, which emanated from the eternal cause of all existence in the beginning of our day of creation, and which has been called the *Logos*, the *Verbum* or *Word*, the *Christ.*

As the Universal One manifested itself, it assumed various aspects, and it therefore appears as a great variety of powers and as innumerable forms of various substances, although all powers and substances are essentially and fundamentally one. The various terms used in occult science are consequently not intended to describe powers and principles radically different from each other, but merely the various aspects of the one universal principle; and as the aspect of things changes according to the point of view from which they are considered, consequently a name applied to a power, if considered from one point of view, may not be applicable if the same principle is considered from another point of view. Likewise, the four sides of a pyramid originate in one point and end in one, each side appearing to have a distinct individuality of its own. The higher we rise towards the summit, the more does this differentiation disappear, and the more does the Unity of all things and their identity with each other become apparent, until all difference is again absolved in the ultimate *One.* He who knows the *One* knows All; he who believes to know many things knows nothing. The One is the starting-point for all occult science.

A & Ω (ALPHA AND OMEGA). — The Beginning and End of all things; *i.e.* the beginning and end of all manifestation of activity and life in the Cosmos; the *Logos* or *Christ.* See *Logos.*

Compare. Rom. ix. 5.— 1 Tim. iii. 16.— 1 John v. 26.— John viii. 58.— John v. 26.— John xiv. 6.— John x. 9.— John xiv. 1.— John x. 30, 38.— John vi. 40, etc.

A'DAM. — Primal man in his aspect as a spiritual power, containing the male and female elements. The spiritual principle, constituting humanity, before it became differentiated in matter and assumed gross material forms.

1 Gen. i. 26.— Ephes. iv. 9.

THE CELESTIAL ADAM. — The divine man-forming power in its original state of purity as an image of the Creator.

1 Gen. i. 27.— Rom. v. 14.

THE TERRESTRIAL ADAM. — Adam after his "fall"; *i.e.* the original man having become the *distorted* image of God by having lost his original purity in consequence of disobedience to the law and desertion of the straight line of the universal divine will. This disobedience is illustrated by the allegory of the "eating of the apple in paradise"; the "snake" which tempted Adam and Eve is the illusion of self, causing man to imagine to be something different from the universal God, and thus creating within him personal desires.

1 Gen. ii. 17.— 1 Gen. iii. 7.— 1 Gen. iii. 10.— Rom. i. 27.— 1 Gen. iii. 16-19.— Luke iv. 6.— John iv. 32.

A'DONAI. — God in his aspect as the *Summum Bonum* in nature; *i.e.* the Light of the Logos having become manifested in nature.

AER. — Air, *Pneuma*, Soul, a universal and invisible principle. See *Elements.*

A'LCHEMY. — The science of guiding the invisible processes of Life for the purpose of attaining certain results on the material, astral or spiritual plane. Alchemy is not only a science, but an *art*, for the power to exercise it must be acquired; a man must first come into possession of certain powers before he can be taught how to employ them; he must know what "Life" is, and learn to control the life-processes within his own organism before he can guide and control such processes in other organisms. *Chemistry* is not *Alchemy*. The former deals with so-called dead substances, the latter with the principle of life. The composition or decomposition of a chemical substance is a *chemical* process; the growth of a tree or an animal, an alchemical process. The highest *Alchemy* is the evolution of a divine and immortal being out of a mortal semi-animal man.

The Song of Solomon describes alchemical processes.

A'NGELS. — Conscious spiritual powers acting within the realm of the Soul, *i.e.* certain individualized spiritual states of the universal consciousness.

2 Sam. xiv., xvii., xx.— Ps. cxciii. 20.— Matt. xv. 31.— Luke xx. 36.— Ps. xxxiv. 7.

A'NIMA. — See *Soul.*

ANIMATO. — Animation. (Alch.) The act of infusing life into a thing or of causing its own latent life-principle to become active. See *Life.*

ANTIMONY. — (Alch.) A symbol representing the element of the *Earth* in its gross material aspect; primordial matter, also represented as the insatiable *Wulf*, the destroyer of forms.

AQUA. — (Alch.) *Water.* See *Elements.*

AQUILA. — (Alch.) *Eagle*, the emblem of *Jupiter*; the symbol of the Spiritual Soul.

A'RCANUM. — (Alch.) Secret. A mystery which is not within everybody's grasp; a certain knowledge which requires a certain amount of development to be comprehended. It also means certain secrets which are not to be divulged to the vulgar, who would be likely to misuse that knowledge.

Matt. vii. 6.

ARCHÆUS. — The great invisible storehouse of Nature, wherein the characters of all things are contained and preserved. To one aspect it represents the *Astral Light;* in another, *Primordial Matter.*

ARGENTUM. — (Alch.). Silver. Symbolized by the *Moon.*

ASTRAL BODY. — A semi-material substance, forming — so to say — the denser parts of the soul, which connect the latter with the physical body. Each thing in which the principle of life exists, from minerals up to man, has an astral body, being the ethereal counterpart of the external visible form.

ASTRAL LIGHT. — The *Light of Nature*. The *Memory*, or universal storehouse of nature, in which the characters of all things that ever existed are preserved. He who can see the images existing in the Astral Light can read the history of all past events, and prophesy the future.

AZOTH. — (Alch.) The universal creative principle of Life.

BABYLON. — Humanity in her unregenerated state, the world of fashion, superficiality, animality and intellectuality without spirituality. The world of superficial Knowledge, self-conceit, and ignorance, living in externals, and being attached to illusions.

Rev. xiv. 8. — Rev. xvi. 19. — Rev. xvii. — Rev. xviii.

BEAST. — (False prophet, Babylonian whore, etc.) Animality, sensuality, and selfishness; but especially *intellectuality without spirituality*, Knowledge without love, scientific ignorance, skepticism, arrogance, materialism, brutality. The *Antichrist, i.e.* false prophets, who are putting man's authority in the place of the universal truth, who degrade religion into sectarianism, and prostitute divine things for selfish purposes, — idolatry, bigotry, superstition, priestcraft, cunning, false logic, etc.

Rev. xvii. — Rev. xviii.

BIBLE. — The "sacred books" of the "Christians," containing a great deal of ancient wisdom clothed in fables and allegories, and describing many occult processes in the shape of personifications of powers and historical events believed to have taken place among the Jews. Some of the events described in those books seem to have actually taken place on the external plane, while others are merely figurative; and it appears to be at present impossible to determine in the Bible the exact line between fiction and history.

BLOOD. — (Alch.) The vehicle for the principle of Life; the seat of the Will.

BODY. — Matter in a certain state of density, exhibiting a form. A body may be visible or invisible, corporeal or ethereal.

Matt. xxii. 30. — 1 Cor. xv. 42, 51. — Phil. iii. 21.

CABALA. — The science which teaches the relations existing between the visible and invisible side of nature; *i.e.* the character of things and their forms in regard to *weight, number*, and *measure*. It is the knowledge of the laws of harmony which exist in the universe.

CAPUT MORTUUM. — (Alch.) Refuse. Dead matter.

CARITAS. — Spiritual Love, benevolence, charity.

CELESTIAL. — A spiritual, divine state; a state of perfection.

CHAOS. — The universal *matrix* or storehouse of nature. — See *Archæus.*

1 Gen. i. 2.

CHIMIA. — Chemistry. Sometimes the term refers to the Chemistry of Life, Alchemy.

CHRIST. — Spiritual consciousness, Life and Light. The divine element in humanity, which if it manifests itself in man, becomes the personal Christ in individual man. "Christ" means therefore an internal spiritual living and conscious power or principle, identical in its nature with the *Logos*, with which the highest spiritual attributes of each human being will become ultimately united, if that human being has developed any such Christlike attributes. This principle is in itself of a threefold nature, but it appears to be useless to speculate about its attributes, as they will be comprehensible only to him who realizes its presence within himself. — See *Logos.*

NOTE. — The misconception of the original meaning of the term "*Christ*" (Kristos) has been the cause of many bloody wars and of the most cruel religious persecutions. Upon such a misconception are still based the claims of certain "Christian" sects. "Christ" originally signifies a universal spiritual principle, the "Crown of the Astral-Light," coexistent from all eternity with the "Father," *i.e.* the Divine source from which it emanated in the beginning. This principle is said to have on many occasions penetrated with its light certain human beings, incarnated itself in them, and thus produced great heroes, reformers, or *Avatars*. Those who cannot rise up to the sublimity of this conception look upon "Christ" as being merely a historical person, who in some incomprehensible manner took upon himself the sins of the world. There have been so many clerical dogmas and misconceptions heaped around this term, that it appears to be impossible to throw any light upon this matter, unless we call to our aid the sacred books of the Hindus and compare the doctrines of *Krishna* with those of Christ. 1 John v. 20. — 1 Tim. vi. 16. — Hos. xiii. 4. — Jer. xliii. — Luke xxiv. 19. — John xii. 44. — Mark ix. 37 — John xiv. 28. — John x. 29. — John xx. 17. — 1 Pet. i. 21. — John i. 4. — John v. 26. — John v. 30. — Matt. xvii. 2. — John xiv. 6, etc.

COAGULATIO. — (Alch.) Coagulation. The act of some fluid or ethereal substance assuming a state of corporeal density.

Canticl. v. 9-14.

COMBINATIO. — (Alch.) Combination. The act of combining certain visible or invisible things.

CONJUNCTO. — (Alch.) Conjunction. The act of two or more things joining together or coming into harmonious relationship with each other.

CORPUS. — (Alch.) Body. Matter is a state of corporeal density. The vehicle of a power.

CREATION. — The external, visible manifestation of an internal, invisible power. The production of a visible form out of invisible, formless substance. The calling into existence of a form.

NOTE. — The term "creation" has often been misrepresented as meaning a creation of something out of nothing; but we know of no passage in the Bible which might justify such an irrational definition. The only persons who believe that something can come from nothing are certain self-styled "scientists," who imagine that life and consciousness are products of the mechanical activity of the body; which is identical with saying that something superior can be produced by something inferior; in other words, by something which according to all known laws of nature is not able to produce it.

CROSS. — A symbol expressing various ideas, but especially the creative power of *Life* in a spiritual aspect, acting within the Macrocosm of nature and within the Microcosm of man. It also represents Spirit and Matter ascending and descending. The perpendicular beam represents Spirit, the horizontal bar the animal or earthly principle, being penetrated by the divine Spirit. Universal as well as individual man may be symbolized by a Cross. Man's animal body is a Cross, or instrument of torture for the soul. By means of his battle with the lower elements of his constitution, his divine nature becomes developed. By means of his physical body, man is *nailed* to the plane of suffering appertaining to terrestrial existence. The animal elements are to die upon that Cross, and the spiritual man is to be resurrected to become united with the Christ. "Death upon the Cross" represents the giving up of one's own personality and the entering into eternal and universal life. The *inscription* sometimes found at the top of the Cross, consisting of the letters I. N. R. I., means, in its esoteric sense, *Igne Natura Renovatus Integra;* that is to say: By the (divine) Fire (of Love) all Nature becomes renewed. The *golden Cross* represents spiritual Life, illuminated by Wisdom. It is the symbol of immortality.

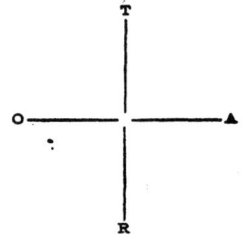

DEUS. — God.

DEVIL. — The principle of Evil, the antithesis of the principle or cause of Good, in the same sense as *Darkness* is the antithesis of Light. *God,* being the cause of all powers and principles, is also the cause of the "Devil," but not its direct cause; for as *evil* is nothing else but perverted *good,* likewise the power called *Devil* is, so to say, the reaction of *God,* or the cause which perverts good into evil. The devil may be said to be the dark, and consequently inferior counterpart of God; consequently, like God, a *Trinity of thought, word,* and its *manifestation.*

Rev. xvii. 8. — Rev. ii. 13. — Luke iv. — 2 Thess. ii. 9-12. — Acts viii. 9. — Mark xiii. 13. — Rev. xvi. 14. — Mark v. 18. — Wisd. xii. 2. — John xxxii. 31. — Eph. vi. 12.

EARTH. — See *Elements.*

EAGLE. — (Alch.) The spiritual Soul. "*The Gluten of the White Eagle,*" — pure spiritual love, the fiery substance of the spiritual Soul.

ELEMENTA. — (Alch.) Elements. Universal and (to us) invisible principles, the causes of all visible phenomena, whether they are of an *earthly* (material), *watery* (liquid), *airy* (gaseous), or *fiery* (ethereal) nature.

There are consequently four " Elements," namely: —

1. *Earth,* representing primordial matter, an invisible ethereal substance, forming the basis of all external corporeal appearances.

2. *Water,* referring to the realm of the Soul, the connecting link between spirit and matter. It also represents Thought.

3. *Fire,* representing the realm of the Spirit or *Life.*

4. *Air,* alluding to Space or Form. It is not, strictly speaking, an "Element."

There is a *fifth element,* which is the spiritual *Quint-essence* (the *Mercury*) of all things. Each element may be considered from a variety of aspects. Each element constitutes, so to say, a world of its own, with its own inhabitants, the "elementary spirits of nature"; and by a combination of those elements under various conditions, an endless variety of forms is produced.

ELOHIM. — The Light of the Logos in its aspect as a spiritual power or influence, whose presence may be felt as it penetrates the soul and body of the worshipper in his moments of spiritual exaltation. This Light, having been the cause and beginning of creation, the term *Elohim* also expresses its aspect as the creative power of the universe.

EVA. — *Eve.* The female or generative power in nature; the eternal mother of all, an ever-immaculate virgin; because she has no connection with any external god, but contains the fructifying spiritual principle (the *Holy Ghost*) within her own self.

The celestial Eve represents *Theo-Sophia,* divine Wisdom, or Nature in her spiritual aspect.

The terrestrial Eve represents Nature in a more material aspect, as the womb or matrix out of which forms are continually evolved, and into which they are reabsorbed.

NOTE. — Primordial man was a bisexual spiritual being; the separation of sex took place in consequence of the differentiation of spirit in matter. Man is still to a certain extent bisexual; because each male human being contains female, and each female being male elements. Sex is merely an attribute of the external form; the spiritual man who inhabits the outward form has no particular sex. 1 Gen. ii. 8. — 1 Gen. i. 27. — Heb. vii. 3. — 1 Gen. vi. 3. — Luke xx. 35.

EVIL. — The antithesis of *Good, i.e.* the reaction of good against itself, or good perverted. There can be no absolute Evil, because such a thing would destroy itself.

EX CENTRO IN CENTRUM. — Everything originates from one centre and returns to that centre.

FAITH. — Spiritual knowledge. A power by which the spirit may feel the existence of truths which transcend external sensual perception. "*Faith*" should never be confounded with "*Belief*"; the latter being merely a controvertible opinion about something of which nothing is known. *Faith* rests upon direct perception; *Belief,* upon intellectual speculation.

Wisd. iii. 4. — Rom. iv. 21. — 2 Tim. i. 12. — 1 John iii. — 2 Cor. iv. 13. — Luke ix. 23. — John xi. 40. — Matt. xxi. 22. — Mark ix. 23. — Matt. xvii. 20, 21. — Luke xvii. 6. — Mark xvi. 17, 18. — John xiv. 12. — Thess. iii. 18. — 2 Cor. xiii. 15.

FATHER. — (Trinity.) The divine and incomprehensible *Fire,* from which emanated the *Light* (the *Son*). We cannot conceive "*the Father*" except as the incomprehensible *Absolute,* the Cause of all existence, the Centre of Life, becoming comprehensible only when he manifests himself as the "Son." In the same sense a geometrical point is merely an abstraction and incomprehensible, and must expand into a circle before it can become an object of our imagination.

1 Cor. viii. 6. — Mark xii. 29-32. — 1 John v. 7. — Wisd. vii. 26. — 1 John i. — 1 John v. 1. — Tim. vi. 16. — John. i. 5. — Eph. i. 23. — 1 Cor. xii. 6. — 1 Tim. vi. 16.

FIAT. — The active expression of the Will and Thought of the *Great First Cause* by which *God* manifested himself in the act of creation; in other words, the energy by which he threw the Light which created the universe into an objective existence. The *outbreathing of Brahm* at the beginning of a *Manvantara. Fiat Lux,* — Let there be Light!

FIDES. — See *Faith.*

FIRE. — An internal activity whose external manifestations are heat and light. This activity differs in character according to the plane on which it manifests itself. "*Fire*" on the spiritual plane represents Love or Hate; on the astral plane it represents Desire and Passion; on the physical plane, Combustion. It is the purifying element, and in a certain aspect identical with "*Life.*"— See *Elements.*

FIRMAMENT. — Realm. Space in its various aspects. The physical and mental horizon. That which limits the physical or mental perception. The sky.

FIXATIO. — (Alch.) Fixation. The act of rendering a volatile substance (for instance a thought) fixed. The act of rendering the impermanent permanent.

Cant. ii. 12.— Cant. viii. 4.

FOUNDATION. — The Real. The basis or centre of things, in contradistinction to their phenomenal illusive and transient appearance. We may look upon all things as having a common basis, which in each thing manifests certain attributes. We may know the attributes of things, but not the thing itself.

GLUTEN. — Adhesion. Spiritual Substance. — See *Eagle.*

GOD. — The eternal, omnipresent, self-existent Cause of all things, in its aspect as the Cause of all Good. The meaning of the term "God" differs according to the standpoint from which we view it; but in its highest meaning it is necessarily beyond the intellectual comprehension of imperfect man; because the imperfect cannot conceive the perfect, nor the finite the infinite. In one aspect everything that exists is God, and nothing can possibly exist which is not God; for it is the One Life, and in it every being has its life and existence. God is the only eternal Reality, unknowable to man; all that we know of him are his manifestations. In one aspect God is looked upon as the spiritual central Sun of the Cosmos, whose rays and substance penetrate the universe with life, light, and power. God being *the Absolute*, cannot have any conceivable relative attributes; because as nothing exists but himself, he stands in relation to no thing, and is therefore non-existent from a relative point of view. We cannot possibly form any conception of the unmanifested *Absolute*; but as soon as the latter becomes manifest, it appears as a Trinity of *Thought*, *Word*, and *Revelation*, *i.e.* as the "*Father,*" the "*Son,*" and the "*Holy Ghost.*"

Innumerable people have been killed because they differed in regard to their opinions how the term "God" should be defined; but it is obvious that a Cause which is beyond all human conception is also beyond any possible correct definition, and that, therefore, all theological disputations about the nature of God are absurd and useless. John iv. 24.— Eph. i. 23.— John x. 26.— 1 Kings viii. 27.— Matt. v. 35.— John i. 18.— Col. i. 15.— 1 Tim. vi. 16.— 1 John i. 5.— Ps. civ. 30.— Wisd. i. 7.— Eph. i. 16.

GOD. — A human being in whom divine powers have become active. An *Adept.*

GOOD. — Everything conducive to a purpose in view is *relatively* good; but only that which leads to permanent happiness is permanent Good. Everything, therefore, which ennobles and elevates mankind may be called good, while that which degrades is evil. Supreme Good is that which establishes real and permanent happiness.

GOLD.— (Alch.) An emblem of perfection upon the terrestrial plane, as the Sun is a symbol of perfection on the superterrestrial plane. There is a considerable amount of historical evidence that the ancient Rosicrucians possessed the power to transmute base metals into gold by alchemical means, by causing it to grow out of its own "seed," and it is claimed that persons possessing such powers exist even to-day.

GRACE. — A spiritual power emanating from the *Logos*. It should not be confounded with "favor" or "partiality." It is a spiritual influence comparable to the light of the sun, which shines everywhere, but for which not all things are equally receptive.

Matt. vii. 16.— 1 Cor. xv. 10.— Rom. xii. 3.— Eph. iv. 7.— John vi. 14. — Matt. xx. 15.— 1 Cor. iii. 6.— John iii. 27.— John vi. 44.— 1 Pet. i. 13-16.— 1 Cor. vii. 7.— 1 Cor. vii. 17.— Matt. xxii. 14.— Matt. xx. 16.— 1 Cor. xii. 31.— Rom. ix. 11, 12.

HEAVEN. — A state of happiness and contentment. Man can only be perfectly happy when he forgets his own self. "Heaven" refers to a spiritual state, free from the bonds of matter.

1 Cor. xv. 50.— Jer. lxvi. 18.— Luke xii. 34.— Luke xvii. 21.

HELL. — The antithesis of Heaven; a state of misery and discontent. A person suffers when he is conscious of his own personality and its imperfections. Each being suffers when it is surrounded by conditions which are not adapted to its welfare; consequently, the soul of man surrounded by evil elements suffers until the elements of evil are expelled from his organization. The state in which the divine and consequently pure spirit is still connected with an impure soul, seeking to throw off the impurities of the latter is called *Purgatory* (Kama loca). When this has taken place, the consciousness of the disembodied entity will be centred in his spiritual organization, and he will be happy; but if the consciousness has been centred in the impure soul, and remains with the latter, the soul will be unhappy and in a state of Hell. The latter takes place especially in such cases where people of great intellectual powers, but with evil tendencies, perform knowingly and purposely evil acts.

Jer. lvii. 21.— Rom. i. 27.— Mark ix. 44.— Rev. xx. 10.— 2 Pet. ii. 17. — Wisd. v. 1-15.

HOLY GHOST. — (Trinity.) The Light of the manifested *Logos*, representing the body and substance of Christ. The Spirit of Truth, coming from the *Father* and *Son*.

John xiv. 17.— John xv. 26.— Rom. i. 20.— Wisd. i. 17.

HOMO. — Man.

HOPE. — Spiritual hope is a state of spiritual consciousness, resulting from the perception of a certain truth, and based upon a conviction that a certain desire will be realized. This kind of hope should not be confused with the hope which rests merely upon opinion, formed by logical conclusions or caused by uncertain promises.

HYLE. — The universal primordial invisible principle of matter, containing the germs of everything that is to come into objective existence. — See *Archæus.*

IGNIS. — *Fire.*

ILLUSION. — All that refers to *Form* and outward appearance. All that is of a *phenomenal* character; transient and impermanent; in contradistinction to the *Real* and Permanent.

JEHOVAH. — *Jod-He-Vah.* — God manifest, in his aspect as the creative, transforming, and regenerating power of the universe. The self-existent, universal God.

JERUSALEM. — Humanity in its spiritual condition. The soul in a state of purity.

JESUS. — The divine man. Each man's spiritual *Ego*. Each person's personal god or *Atman*. The redeeming principle in Man, with which man may hope to become united during his life.

Jesus of Nazareth is believed to have been an Adept; *i.e.* a pure and great man, teacher and reformer, in whom the Logos has taken form; in other words, a human being in whom the Christ-principle has incarnated itself.

John i. 14. — Luke xxiv. 19. — John x. 9. — John xiv. 28. — John i. 4. — John v. 26, 30. — John xiv. 6. — John x. 30, 38.

JUPITER. — The supreme God. Jehovah.

KNOWLEDGE. — Science, based upon the perception and understanding of a truth. It should never be confounded with "*learning*," which means the adoption of certain opinion or theory on the strength of some hearsay or logical speculation. We cannot really know anything except that which we are able to perceive with our external or internal senses.

1 Cor. iii. 19. — 1 Cor. xiii. 8, 9. — Wisd. vii. 23. — Wisd. vii. 17. — 1 Cor. xiv. 1. — Gal. vi. 3. — 1 Cor. i. 20. — Job xxviii. 28. — Wisd. x. 21. — Matt. x. 19. — 1 Cor. i. 19. — Wisd. vii. 13. — Wisd. vi. 13.

LAPIS PHILOSOPHORUM. — (Alch.) A mystery, known only to the practical occultist who has experienced its power.

1 Cor. iii. 16. — Heb. viii. 2. — Matt. xxi. 42. — 1 Pet. ii. 4. — Eph. ii. 21, 22.

LEAD. — (Alch.) symbolized by *Saturn*; the emblem of Matter; the element of Earth.

LEO. — (Alch.) *Lion.* The symbol of strength and fortitude; corresponding to *Mars*. "*The Blood of the Red Lion*," the vehicle of the Life-principle.

LIFE. — A universal principle; a function of the universal Spirit.

NOTE. — Life is present everywhere, in a stone or plant as well as in an animal or man, and there is nothing in nature which is entirely destitute of life; because all things are a manifestation of the *One Life*, which fills the universe. In some bodies the activity of life acts very slow, so that it may be looked at as dormant or latent, in others it acts rapidly; but a form which is deserted by the life-principle ceases to exist as a form. Attraction, Cohesion, Gravitation, etc., are all manifestations of life, while in animals this activity enters a state of self-consciousness, which is perfected in man. To suppose that Life is a product of the mechanical or physiological activity of an organism is to mistake effects for causes, and causes for effects. — See *Creation*.

1 Cor. xv. 53. — John vi. 44. — Luke v. 14. — John i. 4. — Luke vi. 19. — Luke v. 17. — Acts viii. 17.

LIGHT. — An external visible manifestation of an internal invisible power.

The *Divine Light of Grace* is a spiritual Light, the Light of the *Logos*, illuminating the mind of the *Adept*.

The *Light of Nature* in the *Astral Light*.

LIMBUS. — The universal *matrix* of all things. — See *Archæus*.

LOGOS & LOGOI. — A centre or centres of spiritual activity, Life and light, existing from all eternity in the manifested *GOD* (the *Absolute*). The Christ-principle, which, shining into the heart of man, may produce an *Avatar* or *Christ*.

NOTE. — It is taught that at certain periods such an incarnation of the divine Light of the Logos takes place upon the Earth, and thus causes a new saviour, redeemer, and reformer to appear among mankind, teaching the old and half-forgotten truths again by word and example, and thus producing a new revival of the religious sentiment. The ancient religions speak of several such *Avatars* in which "the *Word* has become Flesh." John i. 14. — Col. ii. 19. — John xvi. 27. — Zach. xiii. 1. — 2 Cor. iv. 4. — Heb. vii. 16. — 1 John v. 20. — 1 Tim. vi. 16. — Hos. xiii. 4. — Jer. xliii. 1, 2. — John xii. 44. — Mark ix. 37.

John xiv. 28. — John x. 29. — John xx. 16. — John i. 4. — John v. 30. — Luke v. 17. — Rom. ix. 5. — 1 Tim. iii. 16. — 1 John v. 26. — John viii. 58. — John xiv. 6. — John x. 9. — John x. 30, 38. — John xvi. 27. — Mark xii. 29. — Col. ii. 3. — 1 Cor. vi. 17. — Rom. i. 4.

LOVE. — Spiritual Love is an all-penetrating spiritual power, uniting the higher elements of Humanity into one inseparable whole. It is not led by external sensuous attractions. It is the power by which man recognizes the unity of the All, and the product of that knowledge which springs into existence, when man recognizes the identity of his own spirit with the spirit of every other being. This spiritual Love should never be confounded with sexual desire, parental affection, etc., which are merely sentiments, subject to attraction and change.

1 John iv. 8. — 1 John iv. 13. — 1 Cor. xiii. 7, 8. — Prov. viii. 35. — 1 Cor. xiii. 2.

LUCIFER. — The bearer of 'Light. An angel of Light, possessed of Wisdom. *Lucifer in his fallen state* is Intellectuality without Spirituality; knowledge without the light of wisdom.

LUMEN. — A power emitting Light.

LUNA. — See *Moon*.

LUX. — See *Light*.

MACROCOSM & MICROCOSM. — The great and the little world; the latter being an image or representation of the former, but on a smaller scale. The microcosm of Man resembles the macrocosm of the universe in all his aspects except in external form.

MATRIX. — (Alch.) Womb. The mother wherein a germ, seed, or principle is brought to ripening. Every germ requires a certain appropriate matrix for its development. Minerals, plants, or animals require a matrix in the incipient state of their growth.

MATTER. — An external manifestation of an internal power.

MERCURY. — (Alch.) One of the *Three Substances*. The Astral Light. The principle of Mind. The spiritual quintessence of all things.

METALS. — (Alch.) Certain occult powers. The "metals" of which a man is made and which produce his virtues or vices are more permanent and lasting than the body composed of flesh and blood.

MOON. — (Alch.) A reflection caused by the rays of the Sun. The Intellect, being a reflection of the divine light emanating from the Fire of the heart.

MORTIFICATIO. — (Alch.) Mortification. The art of rendering the lower elements passive, so that the higher ones can become active. The art of dissolving the body, so that the spirit may become free.

MULTIPLICATIO. — (Alch.) Multiplication. Increase. The character is the great multiplicator.

Cant. vi. 7.

NOTE. — Not only is man thus an image of "God," but every part of our organism has the character of the whole impressed upon it, in the same sense as the qualities of a tree are latent in the seed. It is therefore possible for those who can read in the Light of Nature, to know the character, attributes, and history of a thing by examining one of its parts.

MAGIC. — The science and art of employing spiritual powers to obtain certain results. No one can exercise Magic unless he possesses magic powers, and to obtain such powers man must be spiritually developed. "*Magic*" should never be confounded with "*Sorcery*." The former deals with the

Real, the latter deals with *Illusions*. Magic is the culmination of all sciences, and includes them all; but there can be no true science without wisdom, and no wisdom without sanctification.

MAN. — The *real* man is an invisible internal and spiritual power which, in its outward manifestation, appears as a human being.

NOTE. — Man may be looked upon as an individual ray emanating from the great spiritual Sun of the universe, having become polarized in the heart of an incipient human organism, endows the latter with life and stimulates its growth. At a certain state of its development that organism becomes conscious of its existence in the phenomenal world, and with this the illusion of self is created. There is nothing real and permanent about the being called *Man*, except this internal divine power which is called the *Spirit*, which is ultimately identical with the universal Spirit — the *Christ*.

1 Gen. i. 27. — 1 Gen. ii. 7. — Acts xvii. — 1 Cor. iii. 16. — 2 Cor. vi. 16. — Luke iii. 38. — Luke xx. 35.

MARS. — The power which endows beings with strength. — See *Leo*.

MARIA. — The universal matrix of Nature. *Ceres, Tris,* etc. — See *Eve*.

MATERIA PRIMA. — (Alch.) Primordial Matter. *A'Wâsa*. A universal and invisible principle, the basic substance of which all things are formed. By reducing a thing into its *prima materia*, and clothing it with new attributes, it may be transformed into another thing by him who possesses spiritual power and knowledge. There are several states of matter, from primordial down to gross visible matter, and the Alchemists therefore distinguish between *Materia proxima*, *Materia remota*, and *Materia ultima*.

NATURAL, UNNATURAL, SUPERNATURAL. — Relative terms, referring to the relations existing between certain things and certain conditions. Everything in Nature is natural in the *absolute* meaning of this term; but not everything is surrounded by such conditions as according to the laws of its own nature it ought to be surrounded by. Air is natural, but to a fish it is not his natural element; a supernatural being is one who exists in a spiritual condition superior to that of lower beings, and in which gross material beings cannot exist.

NATURE. — The external manifestation of an internal creative power. The whole of nature can be nothing else but a *thought* of God, having been thrown into objectivity by the power of his *Word* and grown into forms according to the law of evolution. "*The nature of a thing*" means the summary of its attributes.

1 Gen. i. 1. — Rom. i. 20. — 1 Cor. xv. 53. — Matt. v. 35. — Mark xiii. 15.

NOTHING. — The antithesis of something. The term nothing is sometimes applied to signify something which is inconceivable and therefore *no thing* to us. *Form* is no *thing*; it is merely a shape, and does not exist in the *Absolute*. If a thought becomes expressed in a form, that which was nothing *to us* becomes something.

OCCULTISM. — The science of things which transcend the ordinary powers of observation. The science of things whose perception requires extraordinary or superior faculties of perception. Everything is occult to us as long as we cannot see it, and with every enlargement of the field of our perception a new and heretofore "occult" world becomes open to our investigation. We may speculate about the Unseen; but we cannot actually know anything about it, unless we can mentally grasp its spirit. — See *Knowledge*.

Sir. i. 16. — Wisd. vii. 21–35. — Jer. ix. 24. — Acts x. and xi. — Jacob i. 5. — 1 Cor. xiii. 8, 9.

OCULUS. — Eye.
OCULUS DIVINUS. — The symbol of spiritual consciousness and knowledge.
OCULUS NATURÆ. — The Astral Light.
OMNIA AB UNO. — "Everything originates from the *One*."
PATER. — Father.
PERFECTIO. — (Alch.) Perfection.
PERSON. — An individual, organized, self-conscious being or principle, capable to think and to will different from other beings or principles. An indivisible unity.
PERSONALITY. — Mask. The sum and substance of the attributes which go to distinguish one individual from others. As one and the same actor may appear in various costumes and masks; likewise one individual spiritual entity may appear successively on the stage of life as various personalities.

NOTE. — To comprehend the doctrine of *Re-incarnation*, it should be remembered that at and after the transformation called "death" only those attributes of a person which have reached a certain degree of spirituality, and are therefore fit to survive, will remain with the individual spirit. When the latter again overshadows a new-born form, it develops a new set of attributes, which go to make up its new personality.

Deut. i. 17. — 2 Chron. xix. 7. — Job xxxiv. 19. — Acts x. 34. — Rom. ii. 2. — Gal. ii. 6. — Eph. vi. 9. — Col. iii. 25. — 1 Pet. i. 17.

PHILOSOPHY. — True "Philosophy" is practical knowledge of causes and effects; but what is to-day called "Philosophy" is a system of speculation based upon logical deductions, or *opinions* arrived at by reasoning from that which we *imagine to know* to the unknown.

Wisd. vii. 21. — 1 Cor. iii. 19. — 1 Cor. xiii. 8, 9.

NOTE. — The fundamental basis upon which our modern philosophy rests is erroneous and illusive, because it rests upon the assumption that man could know something without knowing himself; while, in truth, man can possess no positive knowledge of anything whatever except that which exists within his own self, and he can know nothing about divine things as long as the divinity within himself has not become alive and self-conscious. *Philosophy* without *Theosophy* is, therefore, mere speculation, and frequently leads to error.

PHŒNIX. — (Alch.) A fabulous bird: the symbol of death and regeneration.

PRAYER. — An effort of the will to obtain that which one desires. Prayer on the physical plane consists in acts; prayer on the plane of thought consists in thoughts; prayer on the spiritual plane consists in the act of rising in thought up to the highest, and to become united with it.

Mark vii. 6. — Matt. vi. 7. — Jas. iv. 3. — 2 Thess. iii. 12. — Matt. iv. 2. — 1 Thess. iv. 10. — 1 Pet. iii. 4. — 2 Cor. xii. 4. — Dan. vi. 23. — Luke xviii. 17. — Rom. viii. 26. — 1 John v. 15. — John ix. 31. — 1 Cor. xiv. 14. — Mark xi. 24. — Matt. xxi. 22. — John xv. 5.

PRIMUM MOBILE. — (Alch.) Primordial Motion. The first Life-impulse.
PRINCIPIUM. — Principle, Cause, Beginning of Activity.
PRIMA MATERIA. — See *Materia Prima*.
PROJECTIO. — (Alch.) Projection. The act of endowing a thing with a certain power or quality by means of an occult power whose root is the Will.

Cant. viii. 8.

PUREFACTIO. — (Alch.) Purification.
PUTREFACTIO. — (Alch.) Putrification.

Cant. iii. 1.

RAVEN. — (Alch.) A symbol for a certain occult power.
REBIS. — (Alch.) Refuse. Matter to be remodelled.

REGENERATIO. — (Alch.) Regeneration. The act of being reborn in the spirit. The penetration of the soul and body by the divine heat of love and the light of intelligence, emanating from the divine fire within the heart. The awakening and development of spiritual self-consciousness and self-knowledge.

John iii. 3.—John xvi. 33.—Gal. vi. 15.— 1 Pet. i. 23.— 1 John iii. 9.— 1 John ii. 29.—Wisd. i. 4.— 1 John v. 4.—John iii. 10.—John vi. 27.—Gal. iii. 10.—Thess xi. 19.—Luke xx. 35.—Gal. iv. 19.—John iii. 6.

RESURRECTIO. — (Alch.) Resurrection. Initiation into a higher state of existence. The new life into which the perfected elements of a being enters after the imperfect ones with which they have been amalgamated have been destroyed.

Col. i. 27.—Gal. iv. 5, 6.—Job xix. 25.— 1 Gen. iii. 15.—Rom. v. 15.— 2 Cor. v. 15.— 1 Tim. ii. 3.—Rom. vi. 7.— 2 Cor. iii. 17.— 1 Cor. xv. 35.

ROSE. — (Alch.) The symbol of evolution, and unfolding and beauty.

ROSICRUCIAN. — A person who by the process of spiritual awakening has attained a *practical* knowledge of the secret signification of the *Rose* and the *Cross*. A Hermetic philosopher. A real Theosophist or *Adept*. One who possesses spiritual knowledge and power.

NOTE. — Names have no true meaning if they do not express the true character of a thing. To call a person a Rosicrucian does not make him one, nor does the act of calling a person a Christian make him a Christ. The real Rosicrucian or Mason cannot be made; he must grow to be one by the expansion and unfoldment of the divine power within his own heart. The inattention to this truth is the cause that many churches and secret societies are far from being that which their names express.

SAL. — (Alch.) Salt. Substance. One of the three substances. The Will. Wisdom.

Matt. v. 13. — Luke xiv. 34.

SATURN. — (Alch.) The symbol of the universal principle of matter; the producer and destroyer of forms.

SEED. — (Alch.) A germ, element, or power from which a being may grow. There are germs of Elementals, Minerals, Plants, Animals, Human Beings, and Gods.

Luke xix. 26.— Gal. vi. 7.— 1 Cor. iii. 6.— Mark iv. 26.— Matt. xiii. 23. — 2 Cor. ix. 10.— 1 Cor. iii. 9.—John xv. 5, 6.— Luke vi. 43. — Matt. vii. 16.

SILVER. — (Alch.) An emblem of Intelligence, symbolized by the Moon. Amalgamated with *Mercury* (the Mind) and penetrated by the Fire of divine *Love*, it becomes transformed into the *Gold* of Wisdom.

SOL. — (Alch.) — See *Sun*.

SOL-OM-ON. — The name of the Sun of Wisdom expressed in three languages.

SOLUTIO. — (Alch.) Solution. The act of bringing a thing into a fluid condition.

Corpora non agunt nisi fluida sunt.

SON OF GOD. — One of the three powers constituting the Trinity. The Light, or Christ. The regenerated spiritual man. The celestial Adam. The *Logos*. Only the inner spiritual and divine man is a direct Son of God; the unregenerated man is his indirect descendant. The *Spirit* is the Son of God; the *Soul* is the son of the Sun (astral influences); the *Body* the son of the Earth.

1 Gen. i. 27.— 1 Thess. v. 23.— 1 Cor. iii. 16.— 2 Cor. vi. 16.— Luke iii. 38.— Rom. v. 14.

SOPHIA. — Wisdom.

SOPHIST. — Originally this term meant a "wise man"; but now it means a false reasoner, a skeptical speculator, a person who is cunning but possesses no wisdom; one who judges things not by what they are, but by what he imagines them to be; one who dogmatizes about things which he cannot grasp spiritually; a material scientist, a would-be-wise, an intellectual person without love; one who lives, so to say, in his brain and receives no light from his heart.

Rev. xvi. 14.— Rev. xx. 10.— Jer. xxviii.— 1 John iv. 1.— John x. 1.— 1 Tim. vi. 20.— Matt. xxii. 14.

SOUL. — The semi-material principle connecting matter with spirit. It leads, so to say, an amphibious existence between these two poles of substance, and may ultimately become amalgamated either with one or the other. The Body is the mask of the Soul; the Soul, the body of the Spirit.

Rom. viii. 6.— 2 Cor. iv. 16.— Wisd. iii. 4.— Rom. v. 2.— Matt. xviii. 22.

SPES. — Hope.

SPIRITUS. — Spirit. God in his aspect as an eternal, universal, and invisible principle or power in a state of the greatest purity and perfection. The divine element in Nature. The antithesis of Matter, yet "material" in a transcendental sense. Spiritual substance. A conscious, organized, invisible principle. The Substance or Body of Christ. The term "Spirit" is also used to signify the essence or character of a thing, the sum of the highest attributes or powers.

1 Cor. viii. 6.— Mark xii. 29-32.— 1 John v. 7.— John xiv. 17.— John xv, 26.— Wisd. i. 7.— 1 Pet. i. 10.— Luke xvii. 21.— Gal. ii. 20.— 2 Cor. iv. 2.— Phil. iii. 21.— Rom. xiv. 7.

SPIRITS. — Powers.

NOTE. — The modern usage to apply the term "spirits" to disembodied astral forms and souls of men and animals has originated in the modern misconception of the true nature of man.

SUBLIMATIO. — (Alch.) Sublimation. The rising of a lower state into a higher one. Vices may become sublimated into virtues.

Cant. iii. 6.

SUBSTANCE. — That unknown and invisible something which may manifest itself either as matter or force; in other words, that substratum of all things, which is *energy* in one of its aspects, and *matter* in another.

The Three Substances: Salt, Sulphur, and Mercury represent the trinity of all things. They are the basis of all existence, and in each of these three substances the other two are contained. They form an inseparable Unity in a Trinity, differing, however, in its aspects and manifestations. Consequently, in some things the Salt, in others the Sulphur, and in still others the Mercury is preëminently manifest. They represent *Thought, Word,* and *Form; Body, Soul,* and *Spirit; Earth, Water,* and *Fire; Fire, Light,* and *Heat,* etc. — See *Trinity*.

SULPHUR. — (Alch.) One of the three substances. The principle of Love. The invisible fire.

1 John iv. 8.— Matt. xx. 37.— Eph. v. 2.— 1 John. iv. 13.— Prov. viii. 35. — 1 Cor. xiii. 2.

SUN. — (Alch.) The symbol of Wisdom. The Centre of Power or *Heart* of things. The Sun is a centre of energy and a storehouse of power. Each living being contains within itself a centre of life, which may grow to be a sun.

In the heart of the regenerated, the divine power, stimulated by the Light of the *Logos*, grows into a Sun which illuminates his mind.

The spiritual Sun of Grace. The *Logos* or Christ.

The natural Sun. The centre of all powers contained in our solar system.

NOTE.—The terrestrial sun is the image or reflection of the invisible celestial sun; the former is in the realm of Spirit what the latter is in the realm of Matter; but the latter receives its power from the former.—See *Logos.*

SUPERIUS & INFERIUS.—(Alch.) The *Above* and *Below*, the Internal and External, the Celestial and Terrestrial. Everything *below* has its ethereal counterpart above, and the two act and react upon each other; in fact, they are *one* and merely *appear to be* two.

TARTARUS.—(Alch.) Matter. Residuum. A substance which has been deposited by a fluid, or crystallized out of the latter. The gross elements of the soul.

TERRA.—Earth.

TERRESTRIAL.—An earthly or imperfect state.

THEOLOGY.—A system which teaches the nature and action of divine powers and their relation to Man. Some ancient theologies are the products of certain spiritually developed persons who were capable to perceive and understand spiritual truths, and who laid down the results of their experience in certain systems, and described what they knew, usually in some allegorical forms. *Modern Theology* is a system of speculation based upon the knowledge of external symbols and allegories without any understanding of the true meaning of the latter.

THEOSOPHY.—Supreme Wisdom. The knowledge of divine powers obtained by him who possesses such powers. "*Theosophy*" is therefore identical with *Self-knowledge.*

Wisd. vii. 21.—Wisd. vii. 35.—Wisd. vii. 26, 27.—1 Pet. i. 10.—Wisd. vii. 17-27.—Wisd. viii. 18.—Wisd. x. 21.—Matt. x. 19.—Jer. ix. 24.—1 Cor. i. 29.—Wisd. vi. 13.—Sir. i. 13.

THEOSOPHIST.—A person whose mind is illuminated by the spirit of Divine Wisdom. One who is able to mentally grasp the spirit of a thing, and to understand it. One who has attained a self-knowledge of the divine powers existing in his own organization.

TINCTURA.—(Alch.) Tincture. An ethereal or spiritual substance which, by impregnating another substance, endows (tinctures) the latter with its own properties. If a gross principle is penetrated by a higher one, the former is said to be *tinctured* (colored) by the latter one.

TRINITY.—The All. The whole of the Universe. Everything is a trinity, and Three is the number of *Form*. Every conceivable thing consists of *Matter* and *Motion* in *Space*, and the three are forever one and inseparable. "God" is a trinity, and the Universe being a manifestation of God, every part of the Universe must necessarily be a trinity. Everything is a product of *thought, will,* and *substance* (form); *i.e. Mercury, Sulphur,* and *Salt.*

1 John v. 7.—1 Cor. viii. 6.—1 John i. 5.—1 John iv. 8.—Rom. i. 20.—Wisd. i. 7.

UNIFICATION.—*At-one-ment.* The art of uniting into one. Unification with the eternal One is the only aim and object of all true religion. All things are originally one; they are all states of one universal divine consciousness; they merely *appear to be* different from each other on account of the illusion of *Form*. Differentiation and separation exist merely at the surface of the periphery of the All; the *Centre* is one. To become reunited with the Centre is to enter the *Real*, and to become divine and immortal. After a man has become united with his own higher self, he may become united with *Christ.*

NOTE.—This process of regeneration and unification is taught in all the religions of the East, but—although the whole Christian religion is based upon this truth—it is nevertheless universally misunderstood by modern Christians, who expect to obtain salvation rather through the merit of another than by their own exertion. To understand the process of regeneration and unification requires an understanding of the real nature of man and of his relations to nature; a science which in our modern times is nowhere in Europe taught in schools, because our theologians and scientists are themselves ignorant of the true nature of man, and because mankind finds it easier to accept a belief than to acquire knowledge.

Col. i. 27.—Gal. iv. 5, 6, 19.—Job xix. 25.—1 Cor. xv. 53-55.—1 John iii.—2 Phil. iii. 21.

UNIVERSE.—The Cosmos. The All; beyond which nothing can exist, because there is no "beyond." The whole of the visible universe is a manifestation of the internal invisible divine power called the Spirit of God. It is the substance of God, shaped by his thought into images and thrown into objectivity by an exercise of his Will. Whatever God *thinks*, that he expresses in the *Word*, and what he speaks becomes an *Act*. All this takes place according to *Law*, because God is himself the Law, and does not act against himself.

1 Gen. i. 1.—Wisd. vii. 17.—Rev. xxi. 6.—Rev. xxii. 13.—John xvi. 22.—Rom. i. 29.—Rom. i. 20.—1 Kings vi.-viii.—Matt. v. 35.—Mark xiii. 15.

VENUS.—(Alch.) The principle of Love.

VERBUM.—The *Word*, the A and Ω. The Christ or *Logos*. The expression of a divine thought. The power which emanated in the beginning from the Eternal Centre. The origin of all life.

John i. 18.—Matt. xxvi. 64.—John i. 3.—Sir. xliii. 8.—1 Gen. i.—Eccles. iii. 15.—Ps. xxxiii. 6.—See *Logos.*

VIR.—Man. A human being in whom the male elements are preponderating.

VIRGIN, CELESTIAL.—See *Eve.*

VISIBLE & INVISIBLE.—Relative terms; referring to things which are usually beyond the powers of perception of ordinary man in his normal state. What may be invisible to one may be visible to another.

WATER.—See *Elementa.*

WILL.—The one universal and fundamental power in the universe, from which all other powers take their origin. Fundamentally it is identical with Life. It manifests itself in the lower planes of existence as Attraction, Gravitation, Cohesion; on the higher planes as Life, Will, Spiritual Power, etc., according to the conditions in which it acts. The Will is a function of the universal Spirit of God, and there is no other power in the Universe but the Will of God, acting either consciously or unconsciously, natural or unnatural, if perverted by man. Man can have no will of his own; he is merely enabled to employ the universal will acting in his organization during his earthly existence, and to pervert and misuse it on account of his ignorance with the eternal laws of nature.

Rom. v. 19.—Matt. xxvi. 39.—Heb. x. 7, 36.—Matt. vii. 21.—John v. 30.—John vi. 38.—Heb. x. 9.—Heb. x. 10.—Man can only accumulate will-power by obedience to the law.

WISDOM.—The highest conceivable attribute of the Spirit; conceivable—like all other powers—only by him in whom wisdom has become manifest, and who is thereby rendered wise. Wisdom is not of man's making; he cannot invent, but he can acquire it. The same may be said of all other spiritual powers; they exist in the universe, and are to be attained by Man.

Wisd. vii. 17–27. — Wisd. viii. 18. — Wisd. x. 21. — 2 Gen. iv. 12. — Matt. x. 19. — 1 Cor. i. 19. — Wisd. vii. 13. — Jacob i. 5. — Sir. xxxix. 7, 8. — Wisd. vi. 13. — Wisd. vii. 7. — 1 Pet. i. 10. — 2 Pet. i. 19. — 1 Cor. xiii. 8, 9.

WOMAN. — A human being in whose organization the female elements are preponderating over the male ones.

WORD. — See *Verbum*. A and Ω.

ZODIAC. — The twelve signs of the Zodiac represent the twelve universal principles which form the basis of the construction of the material universe.

SYMBOLS.

THE THREE SUBSTANCES.

♄ Sulphur.   ☿ Mercury.   ⊖ Salt.

THE FOUR ELEMENTS.

△ Fire.   ▽ Water.   ⩟ Air.   ⩡ Earth.

THE SEVEN PRINCIPLES, OR PLANETS.

☉ Sun (Gold).   ♃ Jupiter (Tin).   ☿ Mercury (Quicksilver).   ♂ Mars (Iron).   ♀ Venus (Copper).
☽ Moon (Silver).   ♄ Saturn (Lead).

The Eighth Planet is ☊ the Earth (Antimony).

☌ Day: ☍ Night.

THE TWELVE SIGNS OF THE ZODIAC.

1. ♈ Aries (Ram).   2. ♉ Taurus (Steer).   3. ♊ Gemini (Twins).
4. ♋ Cancer (Crab).   5. ♌ Leo (Lion).   6. ♍ Virgo (Virgin).
7. ♎ Libra (Balance).   8. ♏ Scorpio (Scorpion). 9. ♐ Sagittarius (Archer).
10. ♑ Capricornus (Capricorn). 11. ♒ Aquarius (Waterman). 12. ♓ Pisces (Fishes).

NOTE. — The student should attempt to grasp intuitively the meaning of these signs; for an explanation in words will be entirely inadequate to express their signification unless their spirit is grasped by the power of the intuition.

PART II.

A TREATISE

on the

PHILOSOPHER'S STONE

By a still living Philosopher, but who does not desire to be known.

Written for the instructions of those who love the Secret Doctrine, and for the guidance of the Brothers of the Golden and Rose-Cross.

Copied and Translated from an old German Rosicrucian MS.

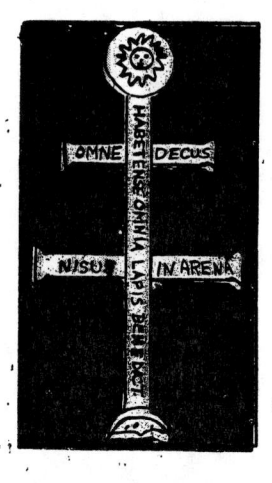

Fig. I.—THE PREPARATION OF THE PHILOSOPHER'S STONE ALLEGORICALLY REPRESENTED.

FIGURE I. represents *Nature*, a great female deity, symbolized as a water-nymph or queen (Venus rising from the ocean), out of which she is born. From her breasts are running two continuous streams of white milk and red blood. The two must be boiled together until they are transformed into silver and gold. Happy is he who surprises the incomparable queen in her secret retirement and obtains possession.

ADDRESSED TO THE DISCIPLE WHO DESIRES TO LEARN THE HERMETIC ART:—

THE reason why, in these last days of the world, I undertook to write this book, is because I wish to prevent you from falling into erroneous opinions. I do not write books for my own sake, for I do not need them. I have read enough books in writing as well as in print during the last twenty years; but I found the greatest part of them filled with phantastry and error. In this book I will describe to you the whole process in the form of a parable, and I will make my description as plain as possible.

As far as my name is concerned, I have determined, after due deliberation, not to reveal it. I do not desire to obtain fame or notoriety before the world; neither do I wish to expose my life to dangers which would unavoidably be connected with the publication of these mysteries, if my person were known. Already some of the Brothers of the Golden Cross have acted imprudently in this respect and were consequently waylaid by certain ambitious and vain fellows, and robbed of their *tincture*,—the highest treasure which a man can possess.

But let me ask those who are unable to understand this book, and who are nevertheless ready to criticise and condemn it: "*Have you seen the great Salt-sea? Tell me, dear, where do they make Sulphur, and where does Mercury come into existence? Have you seen the amorous couple, consisting of man and woman, embracing each other so that eternity cannot separate them, and that they become only one being?*" If you now understand what I mean, if you have seen it with your eyes and touched it with your hands, I will be your brother and ask to be admitted into your laboratory; but if you do not understand it, I advise you to keep your own counsel. Some people complain that this art is very difficult to learn, but let me tell them that those who love God, and are found worthy by him, will learn it very easily; while the godless will never be able to understand it, however much they may exert their own imagination.

If, however, some of you are inclined to accuse me of having described this art too plainly and revealed secrets which ought not to be published, let them know that those who are worthy to learn the art will easily understand what I say, and it will be very useful to them; but those who are not worthy will be sure to leave it alone. I have told the whole process to the would-be wise, and they have laughed at me in their hearts, and could not believe that there is a twofold resurrection of the dead in our work. Our art is a gift of God. He gives it to whomsoever he pleaseth, and takes it away when he chooses, and nobody's own personal will has any influence whatever in determining this matter. It is an art which has been known to me with all its manipulations for over seventeen years; nevertheless I had to wait until, by the grace of God, I could enter into its practice. It is an art which exists as truly as the sun shines during the day and the moon at night, but it exists only for those who are able to see it.

But you, my beloved *Brothers of the Golden and Rosy Cross*, who are keeping yourselves hidden away from the world, and are enjoying the fruits of your divine gifts in secrecy, do not avoid me; and if you do not know who I am, I will tell you that *the Cross is the touchstone of Faith;* it reveals its true value, but imaginary security and sensuality suffocate its germ. Peace be with you all. *Amen.*

PREFACE.

> Quaesivi: inveni: purgavi saepius: atque
> Coniunxi: maturavi: Tinctura secuta est
> Aurea, Naturae centrum quae dicitur: inde
> Tot sensus, tot scripta virum, variaeque figurae.
> Omnibus, ingenue fateor, *Medicina* metallis,
> Infirmisque simul: punctum divinitum ortum.*
>   HARMANNUS DATICHIUS, *Anth. Famulus.*

KIND AND TRUTH-LOVING READER:—

Some years ago, after having long and earnestly prayed to *God*, the unmanifested, incomprehensible cause of all things, I was attracted to *Him*, and by the power of his *Holy Spirit*—through whom all wisdom descends upon us, and who has been sent to us through Christ, the λογος, from the *Father*—he illuminated my inner sight, so that I was able to recognize the *Centrum in Trigono Centri*, which is the only and veritable substance for the preparation of the *Philosopher's Stone*. But although I knew this substance, and had it actually in my possession for over five years, nevertheless I did not know how to obtain from it the *Blood of the Red Lion*, and the *Gluten of the White Eagle*, neither did I know the processes by which these substances could be mixed, bottled, and sealed up, or how they were to be treated by the *secret fire*, a process which requires a great deal of knowledge, prudence, and cautiousness.

I had studied to a great extent the writings, parables, and allegories of various writers, and I had used great efforts to understand their enigmas, many of which were evidently the inventions of their own fancy; but I found at last that all their prescribed methods for the preparation of the *Philosopher's Stone* were nothing but fables. All their *purifications, sublimations, distillations, rectifications,* and *coagulations*, together with their *stoves* and *retorts, crucibles, pots, sand and water baths*, etc., were entirely useless and worthless for my purpose, and I began to realize the wisdom of *Theophrastus Paracelsus*, who said in regard to that *stone*, that it is a great mistake to seek for it in material and external things, and that the people who do so are very foolish, because instead of following Nature, they follow their own brains, which do not know what Nature requires.

Nature in her nobility does not require any artificial methods to produce what she desires. She produces everything out of her own substance, and in that substance we must seek for her. He who deserves her will find her hidden there. But not every one is able to read the book of Nature, and this is a truth which I found out by my own experience; for although the true substance for the preparation of the *Philosopher's Stone* was in my own possession for over five years, nevertheless it was only in the sixth year that I received the key to the mystery by a secret revelation from God.

To open the secrets of Nature a key is required. This key was in the possession of the ancient patriarchs, prophets, and Adepts, but they always kept it hidden away, so that none but the worthy should come into its possession; for if the foolish or evil-disposed were to know the mysteries of nature, a great deal of evil would be the result.

In the following description I have revealed as much of these mysteries as I am permitted to reveal, and I have been strongly forbidden to speak more explicitly and plainly. Those who read these pages merely with their external understanding will obtain very little valuable information; but to those who read them by the light of the true faith, shining from the ever-burning fires upon the altars erected in the sanctuary of their own hearts, the meaning will be plain. They will obtain sweet fruits, and become and remain forever true brothers of the *Golden and Rosy Cross*, and members of our inseparable fraternity.

But to those who desire to know my name, and who might charge me with being too much reserved if I do not reveal it, I will describe it as follows, so that they will have no cause to complain: The number of my name is M.DCXII, and in this number the whole of my name is fully inscribed into the book of Nature by eleven dead and seven living ones. Moreover, the fifth letter is the fifth part of the eighth, and the fifteenth the fifth part of the twelfth. Let this be sufficient for your purpose.

MONTE ABIEGNO, March 25, 1621.

---

* I sought and found; I purified (it) often,
  I mixed (it) and caused (it) to mature.
  The golden tincture was the result; it is called the centre of nature;
  The origin of all thought, and of all the books of men and various figures.
  I now acknowledge freely, it is a panacea
  For all the metals, the weak ones (in the constitution of man),
  And a point which originated from God.

INTRODUCTION.

Fig. II.

FIGURE II. represents the *Prima Materia*, or primordial matter, the foundation of all things, and from which all things are born. It is the dual principle of nature, its parents are the *Sun* and the *Moon*; it produces water and wine, gold and silver, by the blessing of God. If you torture the *Eagle*, the *Lion* will become feeble. The "Eagle's tears" and the "red blood of the Lion" must meet and mingle. The *Eagle* and the *Lion* bathe, eat, and love each other. They will become like the *Salamander*, and become constant in the fire.

THERE have been among many nations, and at various times, certain people who were illuminated by the Spirit of Wisdom, and who wrote books in which they described the result of their knowledge in such a manner that those who earnestly loved the truth for its own sake might be enabled to find it by following their directions. Some of these illuminated people were Egyptians or Chaldeans; others, Greeks, Arabs, Italians, Frenchmen, Englishmen, Hollanders, Spaniards, Germans, Hungarians, Jews, etc., living in widely separated places, and speaking different languages: but nevertheless their writings describe the same process with such a unanimous accord and harmony that the true philosopher may easily recognize the fact of its being the work of only one spirit, speaking through various instruments, with various tongues, and at various times. This harmony exists in all the writings of the sages; but in the books of the wordly-wise we find a great deal of disharmony; for the latter, instead of following the voice of universal truth, which is only one, follow the vagaries of their own brains, which are many, and therefore their opinions disagree and their writings are full of errors.

The writings of the sages differ only in regard to the form in which the truth is expressed, but they all agree in the essential points. They all say that there is only *one* substance, of which the *Philosopher's Stone* is made, and that in this substance is contained everything necessary for its production. This substance is a *spiritual* and *living* one, and all agree, that if you attempt to perform the work with any other substance containing no spirit or life, you will not succeed in your work. Renounce all complexity. Nature is satisfied with only one thing, and he who does not know that thing, cannot command the powers of Nature. This substance is universally distributed everywhere and may be obtained with little expense. It can be found everywhere; every one sees, feels, and loves it, and yet there are only few who know it. *Theophrastus Paracelsus* calls it the *Tinctura Physicorum* or the *Red Lion*; *Hermes Trismegistus* calls it *Mercury*, solidified in its interior; in the *Turba* it is called the *ore*; in the "*Rosarium Philosophrum*" it is called *Salt*. It has as many names as there are objects in the world, and yet it is only known to few. Of this substance may be prepared a spirit as *red as blood*, and another one *white as snow*, and in these two is hidden a third one, the mystery, which is to be revealed by the art. Those who do not know how to begin the alchemical work are yet far from having attained the true knowledge. Those who labor with dead materials will obtain nothing which lives.

Our substance, or *Rebis*, consists of two things, *Spirit* and *Matter;* but the two are only one, and they produce a third, which is the *Universal Panacea*, purifying all things, the *Tincture*, which transmutes base metals into gold. Our *Elixir* is therefore one thing, made of two; but the two are one. The *water* is added to the *body* and dissolves the latter into a *spirit*, and thus the water and body produce a solution. Some philosophers describe the *Philosopher's Stone* as being the true *Spiritus Mercurii* with the *Anima Sulphuris* and the *Spiritual Salt* made into one thing, prepared under one heaven, living in one body; the *Dragon* and the *Eagle;* others call it a preparation made of *spirit*, *body*, and *soul*, and they say that the spirit does not combine with the body except by means of the soul, which connects both together, and yet the three are essentially one.

The omnipotent Creator, whose wisdom extends as far as his (its) own substance, created in the beginning, when nothing but himself existed, two classes of things, the heavenly and the terrestrial. The heavenly things are the interior world, with all its inhabitants; the terrestrial things are the external ones, and have been formed of the four elements. The latter consist of three classes; namely, *Animalia*, *Vegetabilia*, and *Mineralia*, and they are distinct from each other; so that, for instance, the animal kingdom does not produce trees, nor the vegetable kingdom monkeys, etc. But each being has its own peculiar seed by which its own species may be propagated, but no other species is produced by them; the species, however, may be improved, purified, and ennobled to a certain extent, and by appropriate means, as every one knows.

Nature is a great alchemical laboratory in which a continual purification and sublimation to a higher standard takes place. The *primordial matter* from which all the various metals have grown, is originally only one, and contains within itself a *Sulphur*[1] which, acting under various conditions, produced in the course of ages a variety of forms, differing in their exterior qualities, but being essentially only one. Thus a portion of this matter, going through a certain process of evolution, assumed the attributes of iron, and is called Iron; another one became Lead, etc.

Our *philosophical stone* is of a mineral nature, and it is therefore useless to attempt to prepare it from animal or vegetable substances. Nothing can be extracted from a thing unless it is contained therein. Those, therefore, who pretend to be able to make it of such substances are impostors. Moreover, our stone is incombustible, and all animal and vegetable substances are combustible; they will be destroyed in the fire, and nothing remains but smoke and ashes, which are useless for our purpose. Neither can it be made of any imperfect metal or mineral, nor of ordinary Mercury, Sulphur, or Salt, for all these things are destructible in their form.

If you wish to see a thing grow, you must look for its *seed*. A horse is born from a mare, a plant grows by means of the root, and a fire grows out of a spark. If you desire to make gold or silver, you must be in possession of gold and silver; but it must be pure and natural, such as cannot be grasped with the hands. Take such pure *spiritual gold* and sow it into the *white, lamellated earth*, made by a fiery calcination. Cultivate it, and it will grow and bring fruit.

The *philosophical gold* used by the Alchemist is not the common gross material gold, although the latter may be extracted from the former. The gold which he uses is a white and red, true, fixed, and living *tincture*. He uses living gold and living silver; but the ordinary gold and silver are dead and remain dead, no matter to what chemical process they may be subjected. Therefore do not take the dead gold and silver, but take ours, which live.

The beginning of the great work is to dissolve the *Philosopher's Stone* in the water. It is the unification of spirit and body, by which a *mercurial water* is produced. It is a very difficult work, as will be testified to by all who have attempted to perform it; but to him who knows how to prepare this solution, the rest of the mystery will become plain. The solution requires a permanent association of the male and the female elements, from which union a new form may grow.

Let the disciple ponder about the attributes of this water; for the knowledge of the *menstruum* in which the stone is dissolved is the principal condition, without which nothing can be accomplished in this art. It is the great mystery which the sages will not reveal, and which no one is permitted to

---

[1] A power.

tell. The first part of the work is the solution, and consists *in a moistening of the body, so that it may again be dissolved in the substance of the Mercury, and the saltness of the Sulphur be diminished thereby.* This Sulphur is attracted away from two other sulphurs when the spirit meets the body. The second part is the regeneration of the body in the water which is called *Mercury*. There are three Mercuries, the knowledge of which is the key to this science, and without this knowledge nothing will be accomplished in this art. Two of these Mercuries do not belong to the true attributes of the body; the third one is the essential *Mercury* of the *Sun* and the *Moon*. The *Mercury* contained in the other metals is the noblest material for the preparation of the *Philosopher's Stone*.

Our gold and silver are not seen in the rays of the Sun and Moon, but their presence is known by their effects. Our *stone* is the shining substance coming from the Sun and the Moon, by which the *Earth* receives its illumination. No impure matter is added except one, which is the *Green Lion*, and which is used to bind together the two *tinctures* existing between the Sun and the Moon, and to bring them into perfection.

The above remarks will be sufficient for the instruction of the true disciple; but if you do not understand them, then your mind is not yet ripe to know the substance of which the *Philosopher's Stone* is prepared; and you must wait until you arrive at that state by meditation and prayer. I will, however, tell you that the first part of our work is the reduction of the body into its primordial substance; the second is the perfect solution of body and spirit together. The solvent and that which is dissolved remain together, and as the body dissolves, the spirit coagulates.

Thus the whole secret is now revealed to you. If you comprehend it, you will see that it is not at all difficult, but *a work for women and a play for children*, on account of the little amount of labor connected with it. He who knows the beginning knows the end, and his glorified body will behold all the eternal splendor of God, when the illusions of life will vanish before his eyes, being swallowed up in their own insignificance. Follow the teachings contained in the following parable, and the spirit of wisdom will descend upon you and fill you. To arrive at this exalted state we wish to you and to all of us the blessing of the *Father*, the *Son*, and the *Holy Ghost* in all eternity. *Amen.*

PARABLE.

ONCE upon a day I took a walk in a beautiful forest of young and green trees, and began to meditate about the troubles of life, and I remembered how we all became exposed to so much sorrow and misery in consequence of the deep fall of our first parents. I began to weep bitterly. Thus I went on, paying no attention to the place in which I was, and I happened to walk away from the common road. After a while I discovered that I was walking upon a rough and narrow path, all overgrown with weeds, bushes, and shrubs, and I saw that it was a path very seldom used. I attempted to turn back to the main road, but it was impossible to do so; because there was such a strong wind blowing from that direction, that it would have been easier for me to make ten steps forward than one backwards. I therefore made up my mind not to care about the roughness of the road, but to go ahead. I followed that path to a considerable extent, and arrived at last upon a beautiful plain in the form of a circle and grown all over with trees bearing delicious fruits. The inhabitants of that place called it *Pratum felicitatis*. They were all old and venerable-looking men, with long gray beards; all except one, who was quite young, and whose beard was black and pointed; and then there was another and still younger one, whose name I knew, but I could not see his face at that time. They discussed numerous things, and especially a certain great and deep mystery contained in nature, and of which they affirmed that God was keeping it hidden from all the world and revealed it only to those few who loved him very much.

I listened to them for a long time and was very much pleased with what they said, but it seemed to me that the opinions of some of them respecting — not the *materia* or the *work*, — but merely certain forms and allegories, deviated from the ordinary established rules of belief, and that they followed certain methods discovered by Aristoteles, Plinius, and others, and which one copied from the other. I then could no longer restrain myself, but joined in the conversation and refuted their worthless arguments by explaining the results of my own experience. They listened to me very attentively and made me pass through a severe examination; but the foundation of my doctrine was so strong that they could not contradict me. They seemed to be highly astonished and pleased, and unanimously resolved to receive me into their *collegium*, whereat I was exceedingly glad.

But they told me that I could not enter into full fellowship in their society until I was perfectly well acquainted with all the qualities of a certain *lion*,[1] which they had in their possession, and knew exactly his exterior and interior attributes and capabilities. Possessing a great deal of confidence in myself, I promised to try my best to attain that knowledge; for I was so much pleased with their society that I would not have under any circumstances permitted myself to be separated from them.

They accompanied me to the lion and described him to me very exactly; but none of them consented to tell me how to manage him at first. Some of them threw out a few hints about that subject, but they were so dark and uncertain that not one man in a thousand would have been able to understand them. They said, however, that if I would only chain the lion and protect myself against his sharp claws and pointed teeth, then they would explain to me everything. The lion was old, ferocious, and big; his yellow, shaggy mane hung in confused clusters around his neck; he seemed to be invincible, and so ferocious were his looks, that I was at first frightened at my own audacity; but I took courage, for I was ashamed to break my promise, and furthermore these old men stood around, waiting to see

---

[1] The individual will, or personal desire, diverging from the universal will. Man's individual reason is merely a reflex of eternal reason in the individual mind of man; where it usually becomes perverted and tinctured by the desires which exist in the lower (animal) principle. It will then become a strong power, which it is necessary to subdue, before the image of divine reason can be recognized in its original purity.

what I was going to do, and they would have undoubtedly prevented me, if I had attempted to run away. I therefore stepped boldly into the ditch where the lion was kept, and went up to him and attempted to caress him; but he gave me such a ferocious look with his luminous eyes that I became terrified. Nevertheless, remembering that I heard one of the men say that a great many people attempted to subdue the lion and very few succeeded, I did not want to be brought into disgrace, and I thought of a certain trick which I had learned in studying the *athletic art*,[2] and being, moreover, well versed in natural magic,[3] I ceased my caresses and attacked the lion with such boldness and subtility that I succeeded in taking the blood out of his body and even out of his heart, before he became aware of it. I continued my investigations, and found in his body many things which astonished me. His bones were as white as snow, and their quantity was greater than that of his blood.

When the good old men standing around the ditch perceived what I had done, they began to dispute vehemently among themselves. I observed their gesticulations, but I could not understand their words, for I was too far down in that ditch. However, as they became very boisterous, I heard one of them say: "*He must cause that lion to live again, else he cannot become one of us.*"[4] I wanted to act as unceremoniously as possible, so I got out of the ditch, and found myself suddenly upon a big wall, without knowing how I got there. The height of that wall was over a hundred yards, reaching upwards towards the clouds; but on the top it was not more than a foot wide, and there was from the beginning to the end an iron railing, well attached, with many pillars and posts, to the midst of the wall. I say I got upon this wall, and it seemed to me as if some one were walking a few steps ahead of me upon the right side of that railing.

After having followed him for a while, I furthermore perceived that another one was walking behind me on the other side. I do not know whether it was a man or a woman; however, I heard that person call to me, saying that it was easier to walk upon the other side, and I readily believed it; because on account of the narrowness of the road and the multitude of posts and handles which obstructed it, it was very difficult to walk at such a height, and I noticed how several people who attempted to walk behind me, fell down. I therefore grasped the iron railing firmly with both hands and swung myself upon the other side, and continued to walk until I came to a part of the wall which was very steep and dangerous. Then I felt sorry that I had not remained on the other side; but I could neither cross over to that other side, nor was it possible to return. I therefore took courage, trusted the strength of my feet, clung to the railing, and got safely down from the wall. Having progressed somewhat farther, I forgot all the danger through which I had passed; the wall and the railing passed out of my sight, and I knew not what had become of them.

I arrived at a beautiful rose-bush, on which there were some white and red roses growing. I took some of them and put them upon my hat. I then noticed a wall which encircled a large garden, and in the latter there were some young men. Some virgins belonging to those young men, and who were on the outside of the garden, were very anxious to go to them; but they did not want to walk all around the wall, neither did they wish to exert themselves a great deal to arrive at the gate. I felt pity for them, and hastened back in the direction from which I had come, but now upon a level road, and I hurried so fast that I soon arrived at a place where I saw several houses, one of which I supposed to be the gardener's cottage. There I found a great number of people, each of whom had his own chamber. They seemed to be slow people; but two of them worked together very diligently, each of them having his own separate work. It also seemed to me that the work which they performed had been heretofore performed by myself, and that I was therefore well acquainted with it. I now saw that all their labor was filthy and nasty, unreal and fanciful to the extent of each one's peculiarities of education, but having no foundation in nature; and knowing that all their labor was only like smoke which passes away, I did not want to stop with them any longer, but continued my way.

As I walked towards the gate of the garden, some of those people looked at me in a threatening manner, as if they were going to prevent me from entering, while others ridiculed my intention, and said: "See! he wants to get into that garden. We have been here very many years, doing service to the gardener, and we have never succeeded in entering. We will laugh at him to see his disappoint-

---

[2] The art of self-control.   [3] Faith.

[4] After the *self-will* has been killed, it must be made alive again; but now, not in opposition to, but in harmony with, the universal will. It is useless to be merely *passively* good and to have no will at all. To become a useful member of that society, he must become a co-operator with good.

ment." However, I paid no attention to their threats, because I knew more about that garden than they did, although I had never been inside, and I went straight up to the gate which was firmly closed, and in which, with ordinary eyes, there was not even a keyhole to be seen; but I knew intuitively that there the gate could be opened. I therefore applied a small pick-lock[a] which I had brought with me, opened the gate, and entered. I then encountered some more closed doors, but I opened them easily and without any great effort. I now found myself to be in a passage, such as may be seen in a well-constructed house. It was about six feet wide and twenty feet long, with a ceiling overhead, and although the other doors were still closed, I could, nevertheless,—after the first door was open,—look through them into the garden, seeing sufficiently clearly to observe all I desired to see.

In the name of God I walked on in that garden, and after a while I arrived at a little square, each side being about six rods long. It was surrounded by rose-bushes, and the roses therein were very beautiful. There was some rain falling, and the sun shone very bright, and a beautiful rainbow appeared. I was about to turn away from there and to go toward the place where I had seen the virgins to whom I desired to offer my assistance, when I saw that the wall had entirely disappeared, and in its place there was only a low fence of twisted twigs, and near that rose-garden I saw the most beautiful virgin, dressed entirely in white satin, with a most noble-looking youth, clothed in scarlet. Their arms were interlocked, and they were carrying fragrant roses in their hands. I went up to them and asked how they had managed to get across that fence, and the lady answered: "*This, my dearly beloved bridegroom, assisted me in getting over it, and now we will leave the garden and enter our chamber to attend to the duties which our friendship imposes upon us.*" I said: "I am glad that You are obtaining Your wish without troubling myself any further. Nevertheless You will see that I was anxious to serve You, and have walked a long way in a very short time to accomplish that purpose."

I then went on and arrived at a mill, the inside of which was built of stones. There were no meal-boxes, nor any implements for grinding, but through the wall I could see several water-wheels turning in the water. I asked an old miller how the grinding was performed, and he told me that the mill-works were at the other side. At the same time I saw a miller's man walking across the bridge and enter the mill. Following him, I was struck with astonishment, for now the wheels which formerly were to the left of the bridge, were now above it; the water which formerly appeared to be white, was now as black as coal; and the bridge itself seemed merely three fingers wide. Nevertheless I dared to return; I held on to some pieces of wood which were above the bridge, and I got safely across without getting wet. I then asked the old miller how many water-wheels he had, and he answered, ten. This adventure caused me to reflect, and I was anxious to know its significance; but when I found that the old miller was not inclined to satisfy my curiosity, I went away.

Not far from that place there was a high hill paved with stones, and upon that hill there were some of the above-mentioned old men promenading in the warm sunshine. They had written a letter to the Sun, signed by the whole faculty of their college, and they were consulting each other about it. I noticed that the contents of that letter were referring to me. I therefore stepped up to them and asked them what they meant. They said: "We mean that You must keep the woman whom You have recently taken as Your lawful wife; but if You refuse to do so, we shall be obliged to report Your behavior to our sovereign."

I answered: "You need not trouble Yourself about this business, for I love her as much *as if I had been born simultaneously with her*. I have grown up with her, and having taken her, I shall keep her, and nothing shall separate us, not even death itself. I love her passionately and with all my heart."

They then continued: "If so, we shall have no cause for complaint. The bride is equally satisfied. We have her consent, and You must now enter the bonds of matrimony."

"Very well," was my answer.

"Then," said one of them, "the lion will begin to live again, and become stronger and more powerful than before."

I then remembered my previous labor and trouble, and somehow it seemed to me that what they said did not concern me, but some other person well known to me. Just then I saw our bridegroom and his bride again. They were dressed as before, and they approached. They were ready to get married, and I was very glad about it; for I had been terribly afraid that the matter was concerning myself.

---

[a] Knowledge.

But when the bridegroom, in his shining *scarlet-colored* clothes, and the bride, in her *white satin* dress which sent out rays in all directions, came up to these old men, they married them immediately, and I was very much astonished to see that the virgin, although she was said to be the mother of her bridegroom, was still so young that *it seemed as if she had just now been born.*

I do not know what wicked sin they had committed, except that being brother and sister to each other, they were held together by such an ardent love that it was impossible to separate them, and they might perhaps have been accused of incest. However that may be, instead of being put upon the connubial couch, they were sentenced to be put into perpetual prison, there to weep forever, to repent of and to atone for their past misdemeanors. But in consideration of their high birth and noble estate, a prison was selected for them which was perfectly clear and transparent like a crystal globe; and this, moreover, served the purpose of exposing them to the public sight, so that in the future they might not be able to do anything whatever in secret, but that all their actions and omissions might be immediately known to the guard who was watching. Before being put into that prison they were stripped of all their clothing, and jewels, and the ornaments they had worn, and they were then forced to cohabit in that chamber in a state of entire nudity. Nor was any one permitted to enter their prison to serve them, but after having been provided with the necessary food and with some water taken from the above-described mill-pond, the door of their prison was firmly closed and bolted, and the seal of the faculty was put upon the lock; and they ordered me to watch them and to warm their prison, because the winter was approaching. I was ordered to take proper care that they would neither burn nor freeze, and that they could not escape or run away; and I was furthermore informed that if I should neglect my duty, and if damage should be caused thereby, I would be severely punished for it. I felt very uneasy about that affair, and my heart began to sink, for I knew that the duty imposed upon me was of no little importance; and I was also well aware that the *Collegium Sapientiae* was not in the habit of vainly boasting, but that in case of any dereliction of duty on my part, they would certainly execute their threats. However, I could not change this matter, and, moreover, the prison was situated in the interior of a strong tower and surrounded by high breastworks and walls, and it was possible to warm the chamber by means of a moderate but continual fire. I undertook the task, and began in the name of God to warm the chamber, to protect that imprisoned couple against the cold, and lo! very soon a very strange occurrence took place; for as soon as my prisoners began to feel the least degree of the heat, they embraced each other with such a passion that the like of it was never seen before, and they remained in that state until the heart of the bridegroom melted away in the heat of his fervor, and his whole body dissolved in the embrace of his beautiful bride, and fell to pieces. But when the fair bride, whose love to him was not less than his love to her, saw what happened, she began to weep, and she shed so many tears that she — so to say — buried him in those tears, and they flooded the chamber to such an extent that the body of her lover was entirely submerged thereby, and could be seen no more. For a short time she went on weeping in that manner, and her sorrow was so great that she resolved not to live any longer, and thereupon she immediately and voluntarily died. O, how great was my grief and affliction to see those whose welfare had been entrusted to my care — so to say — dissolved and dead before my eyes! I expected my certain perdition, and what I dreaded the most, — and more than the punishment which I expected, — was the ridicule and ignominy, of which I was sure that it would be my lot.

For several days I thought and worried about this matter, and studied how to remedy this misfortune, and at last I remembered how *Medea* restored the dead body of *Æson* to life, and it occurred to me that as she had been able to accomplish that feat, it might perhaps not be beyond my own power to do likewise. I then meditated about the manner in which I would perform that experiment, and I made up my mind to continue to apply a moderate heat until the water would have evaporated, in which case I expected to see again the corpses of our lovers, and thus I hoped to escape from all peril and to obtain profit and praise. I therefore continued to apply heat in the same way as before, for forty consecutive days, when I noticed that the quantity of water in the chamber was gradually decreasing, and the two corpses, which in the meantime had become black as coal, were again visible.

This would have happened sooner if the chamber had been less firmly closed and sealed; but I was not permitted to open the door. I perceived that the water rose and the vapors went up to the clouds; but it fell down again from the ceiling above and could not escape, and this continued until the corpses of our bridegroom and his beloved bride were lying before my eyes in a putrid and rotten state, emitting an exceedingly bad odor. In the meantime the bright sunshine acting upon the moist atmosphere within the chamber produced a very beautiful rainbow with splendid colors, the sight of which made my heart glad. But what pleased me most was the circumstance that I could see the bodies of the two lovers again.

Still, there is no happiness unalloyed with sorrow, and thus my joy was not without grief; for the two lovers were dead, and no life could be perceived in their bodies. I knew, however, that their prison was made of such solid material, and so firmly closed, that their souls and spirit could not escape from it, and I therefore continued my labor without interruption, day and night, imagining that the two would not return to their bodies so long as there was any moisture present. Such really was the case, for I observed that towards evening a great many vapors rose from the Earth, which were produced by the sun in the same manner as the sun attracts vapors from the ocean. Those vapors coagulated during the night-time and formed a beautiful and fructifying dew, which fell down early in the morning and moistened the earth, and bathed the two corpses, so that the latter in the course of time, and from day to day, became more beautiful and white. Their beauty and whiteness increased in proportion as the amount of the moisture became less, until at last, when the air had become clear and beautiful, and all the vapors and dampness had disappeared, the spirit and soul of the bride could not remain any longer in the clear and thin atmosphere; but they entered into the clarified and now glorified body of the queen, who soon felt their presence and immediately began to live again. You may believe me when I tell you that I rejoiced exceedingly about it; especially as I saw that she was wearing such a beautiful dress, such as has never before been seen upon the earth, and she had upon her head a costly crown made of pure diamonds. I saw her rise up, and I heard her say: "*Listen, O ye mortals, and know all ye who have been born of woman, that the* SUPREME *has the power to appoint kings and to dismiss them. He makes men rich or poor, according to his will, he kills and causes to live again. Behold in me a true and living example of his power. I was great, and I became small. Having been humiliated, I became exalted and a queen over many kingdoms; having been killed, I was made alive again. Poor as I am, the great treasures of the wise and powerful are entrusted to my keeping. Therefore power was given to me to render the poor rich, to give mercy to the humble and health to the sick; but in spite of all my power I am not so great as my beloved brother, the great and mighty king, who will be resurrected from the dead.*"

When the queen ceased to speak, the sun began to shine very bright, the day was warmer than before, and the *dog-days* were approaching. Long before that time many beautiful and costly dresses were prepared as wedding gifts for our new queen. They were made of black velvet, ash-colored damask, gray silk, silver-taffeta, snow-white satin, and one which was above all others exceedingly beautiful, and made out of silver-cloth, embroidered with costly pearls and splendidly shining diamonds. Now they made for the young king, likewise, various coats of *incarnate* and yellow, of costly stuffs, and finally a red velvet suit embroidered, and provided with a great number of precious rubies and carbuncles. But the tailors who made these clothes were entirely invisible, and I was very much astonished to see how one dress after another was finished, well knowing that nobody but the groom and his bride had entered that chamber; but what surprised me above all was to see that whenever a new suit of clothes or a new dress was finished, the one which had been made previously disappeared, and nobody knew what became of it.

When the last costly coat was finished, the great, mighty king appeared in all his incomparable splendor and glory, and when he saw that he was imprisoned he spoke very sweetly and courteously, and asked me to open the door and to permit him to come out, saying that great benefit would therefrom result to me. My order was, not to open the door under any circumstances; but the majesty of the king, and the eloquence of his speech, were so great that I opened the door without hesitation. As he stepped out of the chamber he acted with such kindness, amiability, and modesty, that it became evident to me that persons of high standing can have no better ornaments than those virtues. Having been exposed to the heat of the dog-days, he was very thirsty, faint, and tired, and he ordered me to go and fetch him some of the quick-running water from under the wheels of the mill. I did as requested, and having very eagerly drunk the water, he entered the chamber again and ordered me to close the door firmly, so that no one could disturb him, or awaken him from his slumber.

There he rested for a few days, and then asked me to open the door. It then seemed to me that he had become still more beautiful and glorious, and his blood richer. He himself became aware of this, and attributing it to the effect of the water, he asked for more, and having obtained it, he drank a great deal more than before, so that it finally became necessary to make his chamber much larger. After the king had drunk as much as he wanted of this precious fluid, which the ignorant considers to be worthless, he became so beautiful and glorious that I never in all my life saw a more beautiful person, nor any more beautiful actions. He then took me into his kingdom and showed me all the treasures and riches of the world; proving to me thereby that not only did the queen speak the truth, but that there is a great deal more truth still to be described by those who know him. There was no end of gold and carbuncles, of rejuvenating and reconstructing powers, of restitution of health and destruction of disease. But the best and most precious of all was that the people of that country knew their creator. They loved and respected him, and obtained from him wisdom and understanding, and finally, after this temporal splendor, eternal happiness; for the attainment of which we ask the blessing of God the Father, the Son, and the Holy Spirit. *Amen!*

ALLEGORY.[1]

I WAS meditating about the wonderful works of the *Most High*, of the mysteries of *Nature*, and of the fiery and ardent *Love* of Humanity, and I thought of the wheat-harvest, when the son of *Ruben Leae* found upon the field the *Dudaim* that was given by *Lea* to *Rachel*[2] as a reward for having cohabited with the patriarch *Jacob*. My thoughts were very *deep*, and extended to *Moses* who took the *solar calf* which *Aaron* had manufactured, and rendered it *potable* by *burning it with fire, crushing it into powder*, and *sprinkling it upon the water* which he gave to the children of *Israel* to drink[3]. I was wrapt in astonishment, and as I grasped the truth, my eyes were opened like those of the disciples at *Emaus*, who recognized the *Lord* by the manner in which he *broke the bread*. My heart was burning within my breast, and I laid down to meditate further and fell asleep, when lo! the King *Sol-Om-On*[4] appeared to me in my dream with all his glory, riches, and power. He was accompanied by his *wives*[5] and *concubines*,[6] and there were *sixty queens, eighty concubines*, and the number of *virgins*[7] was beyond description. But *one was his darling, his dove, the most beautiful and sweetest of all for his heart*. They were going in a long procession, as it is the custom among the Roman Catholics, and in the line the *Centre* was highly venerated and loved, and the name of the *Centre*[8] was like an *ointment* that has been poured out, and whose fragrance surpassed in sweetness all the spices of the East, and its fiery spirit was a *key* for the door of the *Temple*, and its possessor could enter the Sanctuary and *grasp the horns* at the altar.

After the procession was over, *Solomon* showed me the *Only Centrum in Trigono Centri*. He opened my understanding, and I became aware that *behind me* stood a *naked woman*, having a *bleeding wound in her breast*, from which *blood and water were running*, and *her loins were touching each other like two spangles made by the hands of a Master. Her navel was like a round goblet, always full of a delicious drink; her belly was like two young twin does, her neck like an ivory tower, her eyes like the deep wells at Hessbon near the door Bathrabbin, her nose like a tower upon Lebanon looking towards Damascus. Upon her neck stood her head like the Mount Carmel, and her hair was tied into plaits, falling over her shoulders like the purple cloak of a king.*

But the *clothes*[9] she had stripped off were lying at her feet; they were *disgusting, filthy*, and *poisonous*, and she began to say: "I have stripped off my clothes; how could I be contented to wear them again? I have washed my feet; why should I contaminate them now? *The guardians*[10] *that go about in the town have found me; they have beaten me sore and have taken away my veil.*"

When I heard these words I was terrified by fear and by my own ignorance, and I fell down upon the earth. *Solomon* then bade me arise, and said: "Do not fear. What thou beholdest is *Nature uncovered* and the *greatest mystery* that exists below heaven and upon the earth. *She is as beautiful as Tirce, lovely like Jerusalem, terrible as a forest of spears, and yet a pure and immaculate virgin, out of which Adam was born. Sealed and closed is the hidden entrance to her hut, for she dwells in the garden and sleeps in the double caves of Abraham upon the acre Ephron, and her palace is deep down in the Red Sea, in the crystal grottos and transparent clefts. She is born from the air and has been brought up by the fire. She is a queen of the country; milk and honey are in her breast, her lips are*

---

[1] This allegory contains the fundamental truths of Occultism and Theosophy, and a volume might be written to explain fully its manifold signification. To do so would destroy the original purpose for which it was written; namely, to stimulate self-thought and independent research.
[2] These biblical names have nothing to do with historical *persons*, but refer to occult powers in the universe.
[3] The water (of truth) was too ethereal to be swallowed (and assimilated) by the "Children of Israel" without an addition of "*Matter.*"
[4] The three names of the (spiritual) Sun.   [5] Arts.   [6] Sciences.   [7] Yet undiscovered secrets of Nature.
[8] The *Void*, in which alone Spirit can begin to act.   [9] The external forms.
[10] The priests and scientists who cling to illusions, the "legally appointed" keepers of the (supposed) truth.

dripping sweets, sugar and milk is in her mouth and under her tongue, her clothes are to the wise like odors wafted from Lebanon, but to the ignorant they are an abomination. Rouse thyself, look around, behold all these females, and see whether you can find a single one who can be compared to her."

As he spoke these words he gave a sign, and forthwith all the females present were forced to strip themselves[11] naked. I began my search; but I was unable to decide in favor of any one, for my eyes were kept captive, and I could not tell which one was the most charming.

When *Solomon* observed my weakness he separated all the other females from that naked woman, and said: "Thy thoughts are vain, thy intellect is burnt by the Sun,[12] thy memory is like a black cloud, and therefore thou art not able to decide correctly; but if thou wilt not forfeit thy prospect and trifle away thy opportunity, the *bloody sweat* and the *snow-white tears* of this naked virgin may refresh thy heart, restore thy intellect, and purify thy memory, so that thy eyes may see the *Magnalia* of the *Most High, the height of the uppermost and the depth of the lowest.* The foundation of nature and the power and action of all the elements will be plain to thee, *thy intellect will be of silver* and *thy memory of gold; jewels of all colors*[13] will appear before thy eyes, and thou wilt know how they were born. Thou wilt then be able to separate the good from the evil, the rams from the sheep. Thy life will be rest, but the noise of the jingles of *Aaron* will awaken thee from thy sleep, and the sound of the harp of my father *David* will stir thee up from thy slumber."

This speech of *Solomon* frightened me still more, and I was exceedingly terrified; not only on account of his heart-rending words, but moreover on account of the exceeding beauty and loveliness of that royal woman. The king took me by the hand and led me through a *wine cellar* into a secret but very magnificent hall, in which he refreshed me with flowers and gave me *apples* to eat; and the windows of that apartment were made of clear crystals, through which I looked. *Solomon* then asked me what I saw, and I answered: "I see the same room in which I was a little while ago, and from which I came to this place, and I see all thy royal women to the left and the naked virgin to the right. Her eyes are *redder* than wine, her teeth *whiter* than milk, but the clothes lying at her feet are filthier, blacker, and more disgusting than the creek *Kedron.*"

"Select one of these females for thy sweetheart," said *Solomon.* "I esteem them and my virgin equally. I am much delighted with the loveliness of my ladies, and I am not afraid of their dirty clothes." — And *Solomon* turned around, and began to converse with one of his queens.

There was among the ladies an old governess whose age must have been over a hundred years. She wore a *gray dress* upon her body, and a *black cap* upon her head. The former was beset with *snow-white pearls* and lined on the inside with *red taffety*, and embroidered very artfully with *blue and yellow silk.* Her cloak was of *various Turkish colors* and ornamented with *elevated Indian figures.* This old lady gave me secretly a sign, took me aside, and swore to me that she was the mother of the naked woman; that the latter was a chaste, pure, and mysterious virgin, her own daughter, who until then would never have suffered herself to be seen by any man; and although she had been prostituted often and among all nations, even in the open street, nevertheless *no man had ever seen her naked nor touched her*, for she was the virgin of which the prophet said: "*See! we* have a secretly born son who is different from the rest. See! A virgin has given birth to a child, a virgin who is called Apdorossa, that means secretly, and who does not like to associate with others." — "As this daughter is still unmarried," continued the old woman, "her dower and bridal ornaments are laid under *her feet*, on account of the danger of war, so that they may not be taken away by marauding troops, and she be deprived of her jewels." She further said that I should not be frightened by the stench and the horrid condition of her clothes, but that I should select her daughter above all others for my love and lust; and afterwards she would give me a certain liquid with which I might clean these clothes. She promised me that I should obtain a *fluid salt* and an *incombustible oil* which I might use in my household, and would find it an inexhaustible treasure, and that her right hand would continually caress me, while her left hand would be laid under my head.

I was about to declare categorically what I intended to do, when *Solomon* suddenly turned around, gazed in my face, and said: "*I am the wisest upon the earth; beautiful and delightful is my woman, and the splendor of my queens surpasses the gold of Ophir. The ornaments of my concubines overshadow*

---

[11] They were forced to submit to an investigation of their true merits.   [12] The judgment misled by desires.
[13] The "jewels of all colors" represent certain spiritual states.

the light of the sun, those of my virgins the moon. Heavenly are my ladies, inscrutable my wisdom, unfathomable my intellect." I was very much frightened, and bowed low down, and said: "Behold I have found favor before thee because I am poor. Give me therefore this naked virgin, whom I have selected among all for the continuation of my life. Her clothes are soiled and torn, but I will purify them, and I will love her with all my heart, and she shall be my sister, my bride; because *with one of her eyes*, and with *one of the chains from her neck*, she has taken away my heart and made me passionate, so that I am sick for love."—And forthwith *Solomon* gave her to me,[14] and this act created such a stir and tumult among the females[15] that I awoke, and taking it all for a dream I meditated about it until it was time to arise.

But when I arose, and after I had *prayed*, I saw the clothes of the naked virgin lying on the floor by the side of my bed. She herself, however, was nowhere to be seen, and I began to tremble with fear. My hair stood erect upon my head, and a cold sweat commenced to cover my skin; but I strove to take courage. I recalled the dream in my memory, but my mind was not able to understand the meaning of what I had seen. I therefore did not take the trouble to examine the filthy clothes, neither did I dare to take them away, but I left them undisturbed, and moved my bed into another room. Moreover, the stench that arose from these clothes was so strong that during my sleep my eyes had become poisoned and inflamed,[16] and I was therefore unable to see the *time of grace*, neither could my understanding realize the great wisdom of *Solomon*.

But after these clothes had been lying in my chamber for over five years, and as I did not think that they ever could be of any use to me, I at last resolved to burn them, so as to get rid of their presence, and to put them out of my way. I made up my mind to do this on the very next day; but in the following night the old woman again appeared to me, looked at me very scornfully, and said: "O thou ungrateful wretch! Did I not for the last five years entrust to thee the clothes of my daughter, together with her most precious jewels, and thou never didst attempt to cleanse them, neither didst thou expose them to the sun, so that the moths and worms might be removed? And what is still worse, and still more to be regretted, thou even thinkest of throwing them into the fire! Is it not enough that thou art the cause of my daughter's death, and that she has perished through thee?"

When I heard these words I became very angry, and answered: "I do not understand what you mean. Do you mean that I am a murderer? I never saw your daughter again, nor have I heard of her for the past five years. How could *I* be the cause of her death?"

But the old woman would not listen to me, and said: "My words are true. You have sinned against God, and therefore you could not obtain my daughter, neither could I give you the philosophical fluid[17] I promised to you, and with which you could have cleansed her clothes. *Solomon* gave you my daughter voluntarily and willingly, but you detested her dress, and therefore the *Planet Saturn*,[18] *who is her grandfather*, became angry and changed her again into that which she was before she was born. It is you who have made *Saturn* angry by your disrespect, and you are therefore the cause of her death, putrefaction, and final decomposition; for she is the one of whom the *Senior* says: '*Woe to me! Give me a naked woman! My body was invisible and small, while I had never become a mother, until I had been born a second time*,[19] *and then I gave birth to the powers and virtues of all the roots and herbs, and I became victorious in my essence*,'" etc.

These and other heart-rending complaints fell from her lips, and they appeared to me very strange and unjust. I attempted to suppress my anger as much as possible, but I could not keep from protesting solemnly against her accusations. I told her that I knew absolutely nothing about her daughter, much less of her death and putrefaction. I said that although I had kept her clothes in my room for over five years, nevertheless I had never perceived that they could possibly be of any use, and that I was therefore perfectly innocent before God and man.

My excuses seemed to please the old woman; she looked at me more kindly, and said: "I see by the sincerity of your manner that you are really innocent of any wilful crime, and your innocence shall now be rewarded. I will therefore tell you honestly and in great secrecy, that in consideration of the extraordinary love and affection which my daughter felt for you, she left among her laid-off clothes a

---

[14] The truth will be given to him who *seriously* desires it.
[15] The recognition of the truth is followed by an overthrow of intellectual errors and prejudices.
[16] They had become almost blind to the truth, having been influenced by misconceptions and popular creeds.
[17] Enlightened Reason.   [18] The Life-principle.   [19] Until I had become manifest in the soul.

gray marbled casket for your inheritance. This casket is surrounded with a covering of rough, dirty, and black cloth. Clean it of all the filth and evil odor which still adheres to it on account of its contact with the clothes, and after it is well cleaned, you will require no *key*; for it will easily open, and you will find in it the two following things: *First, a silver case full of splendid and polished diamonds which have been ground with lead; and second, a golden jewel adorned with rubies.* This is the whole amount of the relics of my deceased daughter, and all this she made over to you by her last will and testament before she died, as your inheritance. Take that treasure, treat it according to the rules taught by the *Hermetic art*, purify it secretly, but silently, and with great patience, and preserve it in a warm, moist, vaporous, and transparent secret vault, where it will be protected against cold, wind, hail, lightning, thunder, and other injurious influences, until the time of the wheat-harvest arrives, when you will perceive its great and sublime splendor and rejoice in its possession."

While the old woman spoke she gave me a bottle containing the liquid lye. I then awoke and prayed to God earnestly and fervently that he might open my understanding, so that I might find the treasure-box, which I had seen when it was pointed out to me in my dream. After I had finished my prayer I began to search in the pile of old clothes and found the casket, but the cloth that surrounded it was covered with a hard crust that had grown all around it, and which I vainly attempted to remove, for it would neither be softened by the liquid lye, nor could it be scratched away with iron or steel, and the cleansing fluid did not affect it at all. I finally lost my patience and left it alone, not knowing what else I could do. I suspected that it was bewitched, and I remembered the saying of the prophet, "If you were to clean it with *lye*, and to use a great deal of *soap*, still your vices would be only all the more visible to me."[20]

Again a year passed away in vain speculation. I frequently racked my brain in thinking how I might remove that crust from the casket, but I could find no answer. One day, however, I took a walk in my *garden*[21] for the purpose of driving away my gloomy thoughts, and I sat down upon a *square stone*, and fell asleep. *My body slept, but my heart was awake.*[22] Then the old woman appeared to me again, and asked, "Did you obtain my daughter's inheritance?" I felt very melancholy, and said: "No. I have found the casket, but I cannot remove the crust, for the liquid which you gave me does not appear to soften it."

When the old woman heard me say such a foolish thing, she began to smile, and said: "Do you expect to eat oysters and crawfishes without opening the shell? Is it not always necessary to have them first prepared by *Vulcan*, the ancient and honored cook? I did not tell you to attempt to clean the external crust that surrounds the box, but to purify the casket itself with the cleansing fluid I gave to you, and which originated in that casket. Burn the crust away in the *philosophical fire*, and you will succeed better in your work." — She then gave me a few *glowing coals* tied up in *white tinder*, and taught me how to kindle an artificial philosophical fire, to burn away the crust from the casket. I followed her advice, and immediately there began to blow a *wind from the North* and a *wind from the South*, and *they both blew at the same time through the garden.*

I awoke, and after rubbing my eyes I found the glowing coals wrapped up in white tinder lying at my feet. I took them up in great haste and with great joy, and, praying often, called upon the *Lord*. I studied and *practised* day and night, and thought of the true and excellent motto of the philosopher, who said, "*Ignis et Azoth tibi sufficiunt*,"[23] as truth which is also referred to by *Esdra* in his fourth book, where he says, "*He gave me a goblet full of fire, and his form was like fire; I grew* and *wisdom grew in me.*" And God gave me the *fifth state of perception*, and my spirit entered the *eternal memory*. My mouth was opened and closed no more, and after the fortieth night was over, the two hundred and four books were finished; seventy of them were written for the most wise; they were worthy to be read, and I wrote them upon a *box-tree*.

I worked *silently* and hopefully, according to the instructions that had been revealed to me by the little old woman, until after a long time *my intellect became of silver* and *my memory of gold*, as it had been predicted to me by the King *Solomon;* and after I had very carefully and prudently locked

[20] Vices cannot be eradicated unless they are removed by virtues. As long as evil inclinations exist, the results will be evil; but if the *will* is changed, the desires and inclinations will change. Evil acts are only the outward expressions of evil desires, and the desire is more permanent than the act.

[21] In the interior mind.

[22] My external senses and the perceptible faculties of the astral body were asleep, but the inner *spiritual* perceptions were awake.

[23] The (spiritual) *Fire* and the *Life-principle* are sufficient to accomplish the work.

up the treasure-box and secured it according to the directions received, I found the splendid and glorious *lunar-diamonds* and *solar-rubies*, all of which had originated from *one* casket and from *one* country, and I heard the voice of *Solomon*, saying: —

"*My friend is white and red, and elected among many thousand. His locks are curled, and black as the wings of a raven. His eyes are like the eyes of doves washed with milk by the side of the river, his cheeks are like gardens filled with sanative herbs. His lips are like roses dripping with flowing myrrh, his hands are like turquoises, his body is pure as ebony adorned with sapphires, his legs are like marble pillars based upon golden feet. His stature is like that of the cedars of Lebanon, his breath is sweet and delicious. Such is my friend! my friend! Hold fast to him, O you daughters of Jerusalem, and do not lose him until you have taken him home into your mother's house and brought him into her chamber.*"

When *Solomon* finished speaking, I did not know what to answer him, and therefore kept silent; but I thought of opening the locked-up treasure again, so that I might enjoy peace and remain without molestation. Just then, however, I heard another voice, which said: —

"*I conjure you, O daughters of Jerusalem, by the roes upon the field, not to awaken my lady-love until she herself chooses to awaken, because she is like a closed garden, a hidden spring, a sealed-up fountain. She is the vineyard of Baalhamon, the garden of Engeddi, the mountain of myrrh and incense, the bed, the arm-chair, the crown, the palm and apple tree, the flower of Saron, the sapphire, the turquoise, the wall, the parapet, the well of living water, the princess and the love of Solomon in voluptuousness. She is the best beloved of her mother and has been selected by her. Her head is covered with dew, and her locks with the rain that fell during the night.*"

When I heard this speech and revelation I began to understand the purpose of the sages, and I resolved to leave the hidden treasure untouched until, by the *mercy of God*, by the action of *Nature's nobility*, and by the *work of my hands*, everything will be happily finished.

Shortly after this day, and on the day of the new moon, a *solar eclipse*[24] took place, which produced a terrible effect on all who beheld it. At first the sun appeared with *dark green* and *somewhat mixed colors*, but at last he turned *black as coal*, and the heaven, as well as the earth, became dark. Then the people were very much alarmed, but I was glad in my heart, and thought of the great mercy of God and of the *new birth*, to which Christ's fable of the kernel of wheat refers, and how the seed is decomposed and absorbed by the germ that grows out of it, and that if this did not take place it could not grow and bear fruit. *And an arm reached out through the clouds*, and my body began to tremble, for that arm *held a letter in its hand*, to which were attached *four hanging seals*, and in that letter was written, "*I am black, but I am lovely, O you daughters of Jerusalem. I am like the houses of Kedar, and like the carpets of Solomon. Do not despise me because I am black, for the sun has burned my face.*"

But as soon as *that which was fixed* began to act in *the fluid*, a *rainbow* appeared, and I thought of the covenant which the *Supreme* had made, and of the fidelity of my guide who had instructed me, and the result was that, by the *assistance of the planets and fixed stars*, the sun overcame the darkness, and a bright day appeared over the mountains and valleys. Then all fear and terror was over, and all who had lived to see that day were glad and rejoiced. They praised the *Lord*, and said: "*The winter is gone, the rain has disappeared, the flowers have come forth all over the country; the spring has arrived, and the cry of the turtle-dove is heard in the forest. The fig-tree and the grape-vine have sprouted and send out their fragrance. Therefore let us hasten and catch the foxes, the little foxes which are despoiling our vineyard, so that we may gather ripe grapes, full of artificial wine, and obtain milk and honey, and feast and be filled.*"

And when the day was on the decline, and *the evening appeared*, the entire sky became colored, and the constellation of the *Seven Stars* arose and emitted *yellow rays*. The night ran through its regular course, until the morning was dispersed by the *red* rays of the sun. Then the wise men of the country awoke from their sleep, looked toward the sky, and said: "*Who is she that comes forth like the glory of the morning, beautiful as the moon, excellent like the sun, and without a blemish? The glow of her cheeks is fiery and a flame of the Lord, and many waters cannot extinguish her love, nor would the contents of all the rivers be sufficient to drown her. Therefore we will not desert her, for she is our sister, and although she has become little and has no breasts, we will take her again into her*

---

[24] In the realm of the mind caused by a wave of material thought; by an overshadowing of the sun of intuition by superficial reasoning, false logic, and sophistry

mother's house, into the crystal chamber in which she has dwelled before, so that she may be nourished by the breasts of her mother. Then will she grow and go forth like the tower of David with its parapets and battlements, on which are hanging a thousand shields and weapons for the strong."

And as she came out of her palace, the *daughters* praised her, and the *queens* and *concubines* admired her and extolled her virtue, but I fell down upon my face, thanked God, and praised his holy name.

Thus, O you followers of the truth, the great mystery of the sages and the revelation of the spirit has now been accomplished in all its power and glory. *Theophrastus Paracelsus*, the great monarch in the kingdom of mind, in his *Apocalypsis Hermetis*, says that this mysterious essence is contained in the beginning and the end of the world. In its power rest the elements and the *Fifth Substance;*[25] it adapts the elements and the spirit to each other, and truly overcomes the resistance of the former. It is the sole *Noumen*, a unity, a divine and wonderful activity. *No eye has ever seen, no ear has heard, and into the heart of no mortal has ever penetrated, that which heaven has incorporated into this spirit of truth.* In this mystery alone is the truth, and it has therefore been called the *voice of the truth*. It is the power out of which *Adam* and the old patriarchs, *Abraham, Isaak, and Jacob*, have obtained the *Elixir of Life* and great riches. By the power of this spirit the ancient philosophers have discovered the seven free arts and filled their treasuries with gold. By the power of this spirit did *Noah* build his ark, *Moses* his tabernacle, *Solomon* his temple, and mold the golden vessels that were used therein. *Estras* restored the law by its power, and it is said that its possession enabled *Maria*, the sister of Moses, to be very hospitable to strangers. This spirit was universally used among the ancient prophets; it is an *Universal Panacea* for all diseases of body and mind, the last and the highest mystery of Nature. It is the spirit of *God* that fills the infinite universe, and that moved upon the face of the waters in the beginning. The world cannot conceive of it without the mysterious and gracious inspiration of the *Holy Ghost*, or without the secret instruction of those to whom it is known; but the whole world needs it, and desires it, and cannot exist without it; its value cannot be overestimated by man, and the saints of all ages and of all nations, ever since the beginning of the world, have earnestly desired to obtain it. It rises up into the *seven planets*,[26] lifts the clouds, disperses fogs, gives to all things their light, transforms everything it touches into silver and gold, produces health and superabundance, virtues and treasures, comforts the sufferer, heals the sick, and cures all diseases. It is the mystery of all mysteries, the secret of all secrets, the strength and the life of everything. It nourishes the body, refreshes the soul, keeps man in continual youth, drives away old age and debility, destroys weakness, and rejuvenates the world. Its quality is inscrutable, its power infinite, its action invisible, its magnificence greater than all.

It is above all earthly and heavenly things, a spirit of spirits, a select essence, which gives health, happiness, joy, gladness, peace, and love. It destroys poverty, ignorance, and misery. It changes men into beings who can neither think nor speak evil, nor act wrongly, but who are all-powerful for good. It gives to every one that which *his heart* desires. To the good it gives honor and long life; but to the bad, who misuse it, eternal punishment.

And now we will—in the name of the *Holy Trinity*, and in the following words—close that which we had to say about this *great mystery*, the secret of the *Philosopher's Stone*, and we now, thereby, most solemnly celebrate and conclude *the highest feast of the sages:*—

Praise, honor, and thanks be given forever to the most high and all-powerful *God*, who has created this art, and whom it has pleased to reveal it to me through a sacred covenant, according to his promise which he fulfilled in spite of my own imperfections. I pray to him with all the aspirations of my heart, and in great humility of mind, that he may rule and guide my soul, my sense, and my understanding, by the power of his spirit of sanctity, *so that I may not speak of this secret* before the world, much less communicate it to the unworthy, or reveal it to any creature, and thus *break my oath, tear the seal of divinity, and become a perjured brother of the golden Cross;* as by doing so I would heap the blackest of insult upon the majesty of the *Supreme*, and would knowingly and infallibly commit a sin against the *Holy Ghost*,[27] an evil from which *God* as father, son, and spirit, the *Adorable Trinity in Unity*, may protect me forever and ever. *Amen! Amen! Amen!*

---

[25] "*Mercury*," the Universal mind-essence.   [26] The Seven Principles.
[27] The "*Sin against the Holy Ghost*" is the wilful rejection of the Truth after it has once been fully recognized and understood.

PART II.

THE CELESTIAL VIRGIN.
The celestial and terrestrial EVE, the Mother of all Beings in Heaven and on the Earth.
The Star of the Sages from the East.

God is the one, uncreated, infinite, self-existent and eternal SPIRIT, the only REALITY. God in Nature and Time has become a corporeal, visible, mortal MAN.

Nature is a created (formed), temporal, spiritual, corporeal and real POWER or SUBSTANCE, an image, picture, shadow of the uncreated, infinite, eternal Spirit, visible and also invisible.

The Sun of Justice.

OCULUS DIVINUS
per quem Deus vidit et creavit omnia.

OCULUS NATURÆ
sive Coeli, per quem Natura visitat et regit terrena omnia.

The Beginning of a thing is the beginning of its end.

The Virgin Sophia.

Living, dying, transforming, regenerating.

Lumen Gratiæ, Ergon
sunt duo

Lumen Naturæ, Parergon
Fratres.

Celestial Eve,
Regeneration.

Terrestrial Eve,
Reincarnation.

God (the WORD) has become Man.

The Microcosm of Nature produced the Microcosm of Man.

MICROCOSM. — KNOW THYSELF
When thou shalt thou know — knowesth thyself, this symbol.

Tinctura Celestis.

Tinctura Physica.

ROSA CRUCIS
VENITE.
Videte. Videte. Videte
He who has eyes will see.

Seek the friendship of Archao, the trusted Guardian of the Threshold.

PHILOSOPHORUM
VENITE.
Arrigite, Arrigite, aures
He who has ears, will hear.

He is the pledged vassal of Nature, the silent worker in her laboratory, and servant of her privy-chamber

7 Eagles. 7 Lions. 7 Ravens. 7 Spheres.

Sub umbra alarum tuarum.
P. F.
Consummatum est.

The miraculous bird Phoenix, having three eggs. In the first one is air, in the second two yolks, in the third one a young cock.

THORIA — PRACTICA

Multi sunt vocati, SOPHISTA. SYLEX MAGUS. pauci vero electi.

Dominus providebit. Exitus acta probabit.

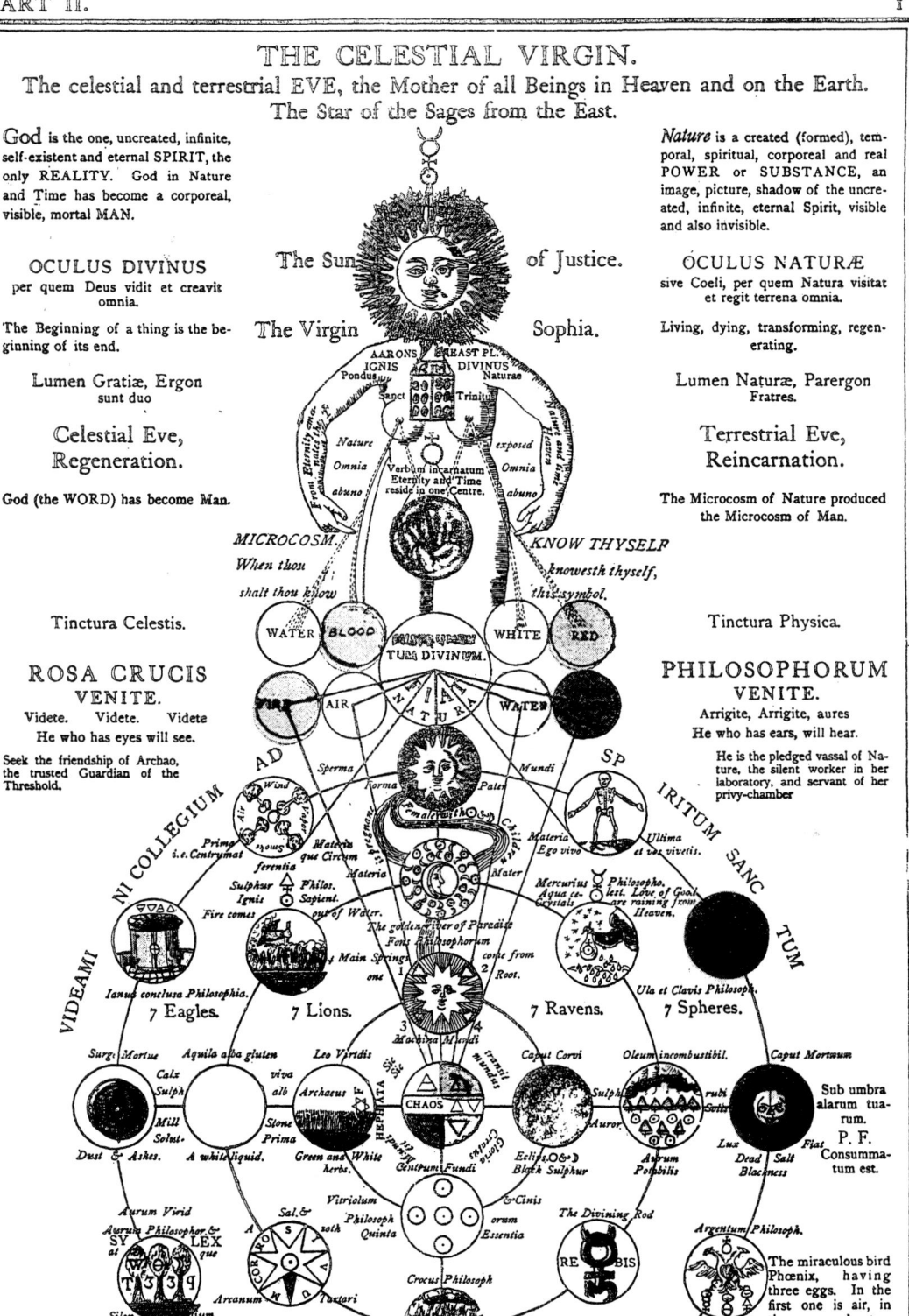

Unification.

Eat, O friends, drink, yea drink abundantly.—*Catn. v., 1.*

He who eats my flesh and drinks my blood, is in me, and I in him.—*John vi., 56.*

The true communion is the substantial, powerful and omnipotent presence of the Christ.—*Taulerus.*

If we are penetrated and full of the spirit of Christ, then is Christ present within us, and we are in Christ.—*Taulerus.*

God is the Light, and there is no Darkness in him.

If we walk in the Light, as he is in the Light, we have fellowship one with another.—*I. John, 7.*

EGREDERE.
per Viam CRUCIS.

INGREDERE.
per Vitam LUCIS.

To him that overcometh, I will give him a white Stone, and in the stone a new name written, which no man knoweth, saving he that received it.—*Revelat. ii., 17.*

CHRIST is the tree and trunk of Life, we are the branches, in which his power is active, producing flowers and fruits in and through us. His soul is the divine quality of the human soul, and if the latter becomes pervaded and penetrated by this divine principle, she becomes through the power and love of Christ divine herself, and is thereby restored to her former state of being, a divine and eternal image and a participator in eternal life. We absorb the substance of the Christ principle within our soul; the real inner man, who is hidden below the gross material animal mask, is a spiritual being, and requires spiritual — not corporeal — nutriment. The union takes place through the power of (spiritual) Faith. The true living faith in Man is itself Christ, who remains in man and is his (spiritual) Life and Light.

Thus we become purified through the power of the living Faith, and illuminated and pervaded by the Light of the Holy Ghost. The elementary (animal) body requires elementary food; the Soul requires spiritual food. Each principle requires that kind of food to which it is in relation, each eats and drinks of that fountain, from whose centre it has been produced. Darkness requires darkness to grow, and Light is fed by Light. They have no communion with each other. The natural body receives its nutriment from the Earth; the sidereal and perishable body from the Firmament (Astral plane); but the Spirit of God is the nutriment of the human soul, and supplies the latter with immortal life, with love, sanctity, tranquillity and happiness.

The Body of Christ is a universal spiritual substance or principle, which fills him who rises up to it in his soul by holy aspiration, and remains forever inaccessible to him who is not worthy to receive it.

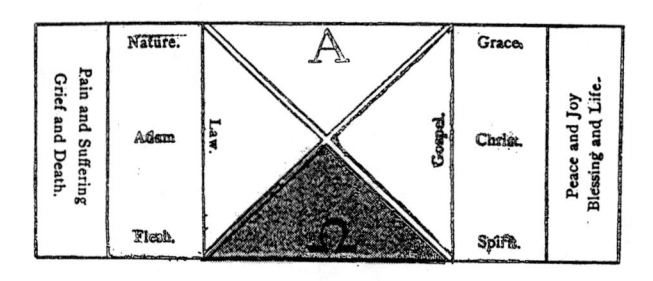

PART II.

FIGURA CABALISTICA
1. 2. 3. 4.

Fire and Light was the Beginning.—*Genes. i. 3.*
Fire will be the End.—*2 Peter iii. 10, 12.*

Fire and Light existed.—*2 Cor. iv. 6.*
Fire and Light.—*1 John i. 6, 7.*
—*1 Tim. vi. 16.*

THE CELESTIAL SUN.
With his Rainbow and 4 colors.
NATURA.

Supernatural and Natural Fire.

The Divine, Incomprehensible Fire-Light In the internal Man
2 Cor. iv. 6. 1 John i. 5.

THE TERRESTRIAL SUN.
With his Rainbow and 4 colors.
TINCTURA:

Invisible and Visible Fire of the four Elements.

Apoc. xxii. 5.

1. Red Color.
2. Yellow Color.
3. Green Color.
4. Purple Color.

☉ and ☽ must become obscured and black. Mortificatio.

1. Degree, Natural Fire.
2. Supernatural Fire.
3. Antinatural Fire.
4. Unnatural Fire.

Nature must become natural, in, by and through herself.

Black Color. 1.
Green Color. 2.
White Color. 3.
Red Color. 4.

Preparat ☉ ☽ dry method
Solve and ⊕ Coagula
Corruptio ⊠ Regeneratio
Completing ⊙⊙⊙ Labour, moist method

Walk on the right and it will guide thee to the natural spiritual life.

Of him, by him and in him are all things.—*Rom. ii. 36.*
In him we live and move and have our being.—*Acts xvii. 28.*

One God and Father of All, who is above all and through all and in you all.—*Ephes. iv. 6.*
God is All in All.—*1 Cor. 15.*
In Eternity and in Time, in Heaven and in Hell.

What will the Sun and the Moon, Fire and Light benefit you, if you remain in Darkness?

What will the truth benefit you if you cling to error?

The world is steeped in darkness and seeks only the eternal light.

Those who seek to know eternal Nature, while they refuse to see the internal and eternal power acting therein, will never know Nature's mysteries.

Thou wilt light my candle. The Lord, my God, will enlighten my darkness.—*Psalm xviii. 28.*

THE SEPTENARY MYSTERIES.

MAN KNOW THY SELF, AND THOU SHALT KNOW ALL.

In the sacred number 1. 2. 3. 4. 5. 6. 7. many divine and natural mysteries are contained. To see them in the Light of the Spirit, and in the Light of Nature, is the true Knowledge of God and Man, and in this is contained all terrestrial and celestial wisdom in Heaven and upon the Earth.

Study the allegories in which these numbers occur in the Bible. 666 see *Revelat. xiii, 18. & xx. 4–10.*
999 + 1. = 1000, the inscrutable △ 999 ♁ h. e. 666 ✡

THEORY. This is the BEAST, the Dragon, false Prophet, the Whore of Babylon, according to its eternal aspect in A.

and the Beast, Dragon, etc., in its temporal aspect in the ♌. **PRACTICE.**

This is REGENERATION and UNIFICATION with GOD.

From one Being come all things in Heaven and on the Earth; Eternity generates Time, and by the Power of the WORD Nature and Time are again swallowed up in Eternity; but the new Being remains in Christ, for it is of a divine nature.

THE SEALED BOOK.

Apocalypses V.

This is the manifestation and testimony; the true recognition of J. C. God and Man; the living Book of Life, containing all celestial and terrestrial Wisdom in Heaven and upon the Earth, the Sealed Book in time and eternity.

See Apocalyps.—*Chaps. V., XX., XXI., XXII.*

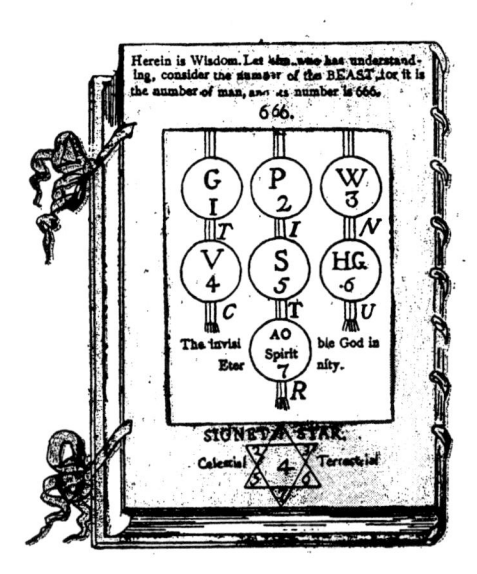

Herein is Wisdom. Let him, who has understanding, consider the number of the BEAST, for it is the number of man, and its number is 666.

666.

G I T	P 2 I	W 3 N
V 4 C	S 5 T	H.G. 6 U

AO Spirit 7 R

The invisible Eternal Spirit AO God in Unity.

SIGNET A STAR.

Celestial 4 Terrestrial

Apocalyps *XLII.*

M N
E BEAST N
E 666 N
E M

$$\text{All} \begin{cases} 1.\ Wisdom \text{ is contained in} & One\ Book \\ 2.\ Power \text{ is contained in} & One\ Stone \\ 3.\ Beauty \text{ is contained in} & One\ Flower \\ 4.\ Riches \text{ are contained in} & One\ Treasure \\ 5.\ Happiness \text{ is contained in} & One\ Good \end{cases} \text{is}$$

called

JESUS CHRISTUS

α † ω

Crucifixus & Resuscitatus.

This is

The Fountain, the Tree, the Light, the true Book of Life and of the Lamb.

He who possesses it

Knows all things in Heaven, upon the Earth and below the Earth; it is the source of all knowledge.

(Intellectuality without Spirituality is the *Sign of the Beast.*)

THE OPEN BOOK

With its Seven Seals.

Apocal. X.

This is the revelation and testimony, the true recognition of J. C., God and Man; the Living Book of Life, containing all celestial and terrestrial Wisdom in Heaven and upon the Earth, and the OPEN BOOK according to His WORD and His Divine Humanity in the World and in time, given by God to His servants.

See *Apocalyps, x., ii, iii, xiii.*

The 1st Cord.
The Lamb with two horns, *i. e.*, the celestial and terrestrial kingdom.

The 2d Cord.
The united two natures, Divinity and Humanity in the person J. C.

The celestial Trinity in its celestial essence, in the Spirit, and invisible God

The terrestrial holy Trinity in Time, Word and Flesh, and in its state of degradation a visible, corporeal God.

Eternity becomes Time

God 1 — Word 3 — Holy Ghost 6

The Beast and its Number 666

Divine Signet Star Eternal Paradise

Philosophical Signet Star Terrestrial Paradise

Happy is he, who knows the number of the Beast. 666. *Ap. XIII.*
The Beast has been, is not and yet is, and will remain in Eternity. *Ap. XVII.*

The 3d Cord.
Generation,
the Old Testament
LAW

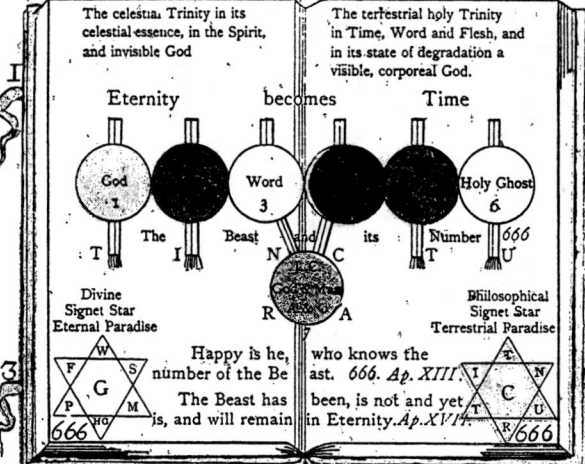

The 4th Cord.
Regeneration,
the New Testament
LOVE.

F S H.G.

☿ ☿

A O
✝
J C

i. e.
Signet Star
Candle-Stick.

7 7
7
7 7
7
Seal

Close the Book again and tie the cords, and it will be again the SEALED BOOK. Then will the holy Trinity of the three celestial, spiritual, invisible and eternal persons, be united and manifest in the terrestrial, temporal, visible holy Trinity, the one person and humanity, J. C., God and Man; One in Heaven and One upon the Earth, All in One and inseparable. In Him (in humanity as a whole) resides the fulness and plenitude of Divinity.—*John, ix, x, etc.*

IESUS CHRISTUS

The Basis and Living Corner-stone.

To know him in the Spirit and Truth is to possess the
Eternal Life.
To Love
DIVINITY in HUMANITY
is better
than all worldly learning.

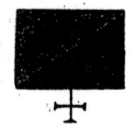

EVOLUTION.

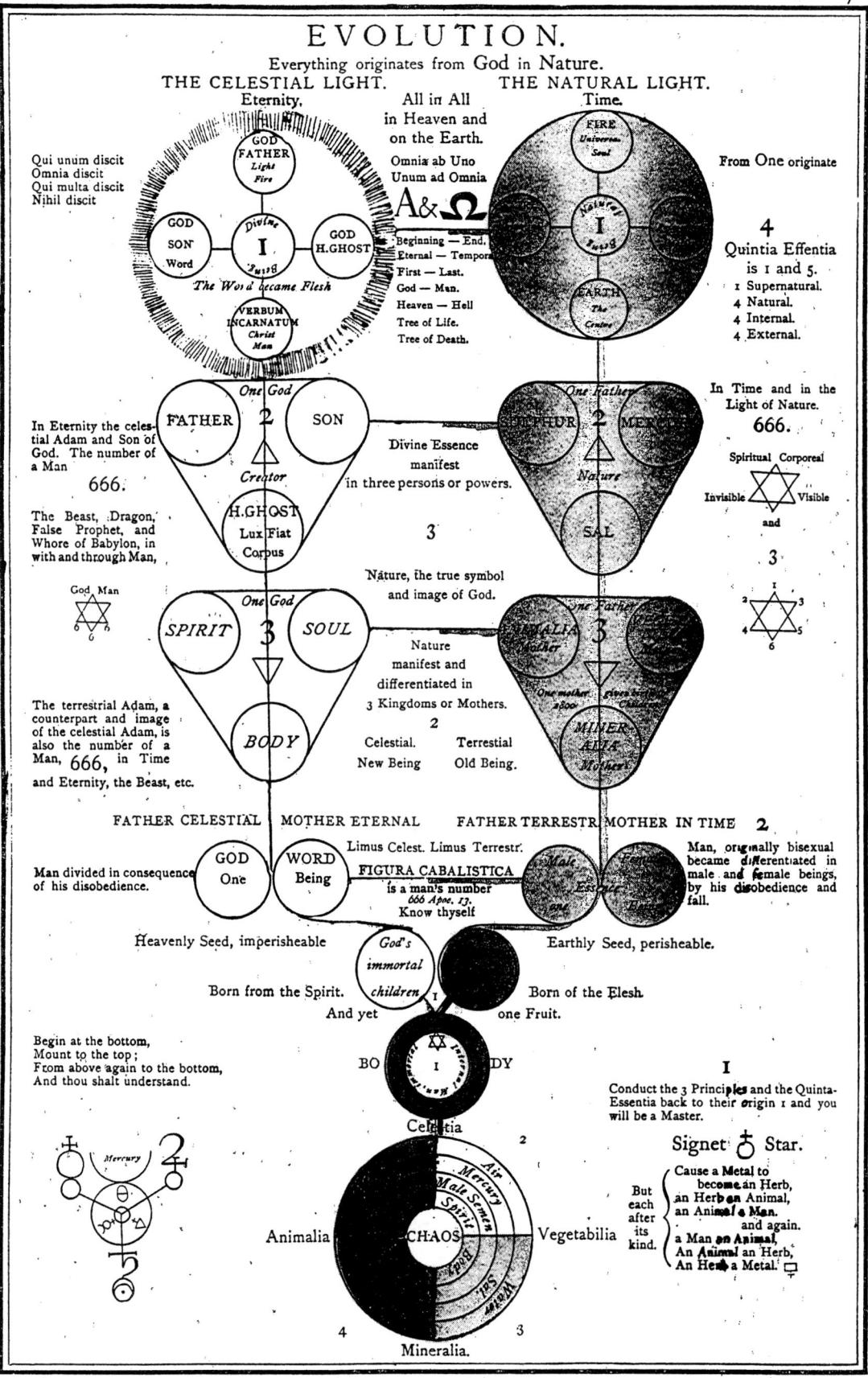

Everything originates from God in Nature.

THE CELESTIAL LIGHT.   All in All   **THE NATURAL LIGHT.**
Eternity,   in Heaven and   Time.
on the Earth.
Omnia ab Uno
Unum ad Omnia

Qui unum discit
Omnia discit
Qui multa discit
Nihil discit

A & Ω

Beginning — End.
Eternal — Temporal.
First — Last.
God — Man.
Heaven — Hell.
Tree of Life.
Tree of Death.

From One originate

4
Quintia Essentia
is 1 and 5.
1 Supernatural.
4 Natural.
4 Internal.
4 External.

In Eternity the celestial Adam and Son of God. The number of a Man

666.

The Beast, Dragon, False Prophet, and Whore of Babylon, in with and through Man,

God Man

Divine Essence manifest in three persons or powers.

In Time and in the Light of Nature.
666.

Spiritual   Corporeal
Invisible   Visible
and
3.

3

Nature, the true symbol and image of God.

The terrestrial Adam, a counterpart and image of the celestial Adam, is also the number of a Man, **666**, in Time and Eternity, the Beast, etc.

Nature manifest and differentiated in 3 Kingdoms or Mothers.

2
Celestial.   Terrestrial.
New Being.   Old Being.

FATHER CELESTIAL   MOTHER ETERNAL   FATHER TERRESTR.   MOTHER IN TIME   2

Man divided in consequence of his disobedience.

Limus Celest. Limus Terrestr.
FIGURA CABALISTICA
is a man's number
666 Apoc. 13.
Know thyself

Man, originally bisexual became differentiated in male and female beings, by his disobedience and fall.

Heavenly Seed, imperisheable.   Earthly Seed, perisheable.

Born from the Spirit.   God's immortal children   Born of the Flesh.
And yet   one Fruit.

Begin at the bottom,
Mount to the top;
From above again to the bottom,
And thou shalt understand.

I
Conduct the 3 Principles and the Quinta-Essentia back to their origin 1 and you will be a Master.

Signet ☿ Star.

But each after its kind.
{ Cause a Metal to become an Herb, an Herb an Animal, an Animal a Man.
and again.
a Man an Animal,
An Animal an Herb,
An Herb a Metal. }

Animalia   CHAOS   Vegetabilia

Mineralia.

THE RIVER OF GOLD AND OF SILVER.

In the following two sentences is contained everything that is hidden within the celestial and natural Light, and he who understands the true meaning of these sentences in their eternal and temporal aspects is a true Theosophist, Cabalist, Magician, and Alchemist. He who can explain them, according to the A. & Ω is a Master and a true Brother of the R. C.

KEEP THIS IN MIND!

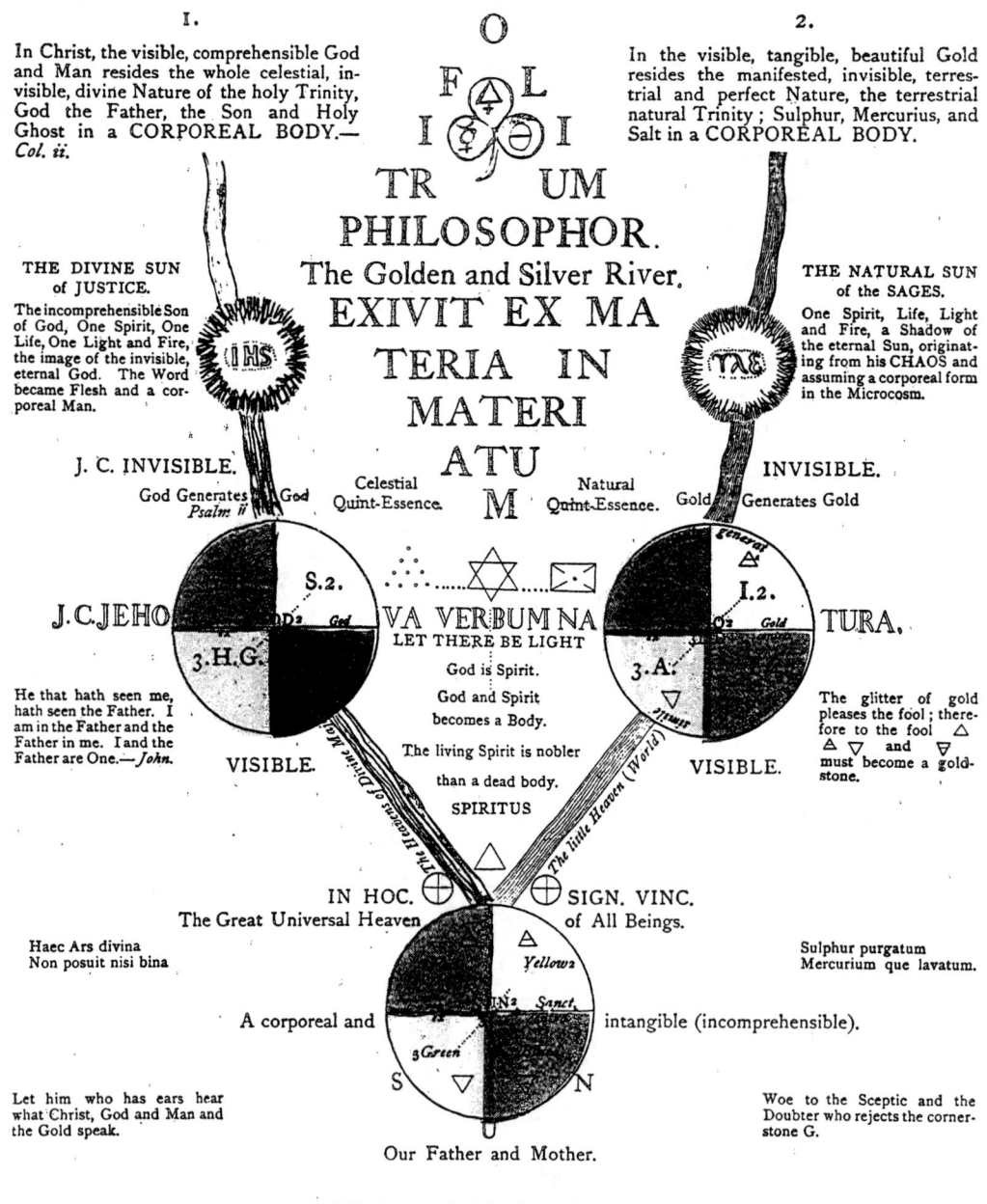

1.
In Christ, the visible, comprehensible God and Man resides the whole celestial, invisible, divine Nature of the holy Trinity, God the Father, the Son and Holy Ghost in a CORPOREAL BODY.—*Col. ii.*

2.
In the visible, tangible, beautiful Gold resides the manifested, invisible, terrestrial and perfect Nature, the terrestrial natural Trinity; Sulphur, Mercurius, and Salt in a CORPOREAL BODY.

TR UM
PHILOSOPHOR.
The Golden and Silver River.
EXIVIT EX MA
TERIA IN
MATERI
ATU
M

THE DIVINE SUN of JUSTICE.
The incomprehensible Son of God, One Spirit, One Life, One Light and Fire, the image of the invisible, eternal God. The Word became Flesh and a corporeal Man.

THE NATURAL SUN of the SAGES.
One Spirit, Life, Light and Fire, a Shadow of the eternal Sun, originating from his CHAOS and assuming a corporeal form in the Microcosm.

J. C. INVISIBLE.   INVISIBLE.
God Generates God   Gold Generates Gold
Psalm ii
Celestial Quint-Essence.   Natural Quint-Essence.

J.C.JEHO   TURA,
VA VERBUM NA
LET THERE BE LIGHT
God is Spirit.
God and Spirit becomes a Body.
The living Spirit is nobler than a dead body.
SPIRITUS

He that hath seen me, hath seen the Father. I am in the Father and the Father in me. I and the Father are One.—*John.*

VISIBLE.   VISIBLE.

The glitter of gold pleases the fool; therefore to the fool △ ▽ and ▽ must become a goldstone.

IN HOC.   SIGN. VINC.
The Great Universal Heaven   of All Beings.

Haec Ars divina
Non posuit nisi bina

Sulphur purgatum
Mercurium que lavatum.

A corporeal and   intangible (incomprehensible).

Let him who has ears hear what Christ, God and Man and the Gold speak.

Woe to the Sceptic and the Doubter who rejects the cornerstone G.

Our Father and Mother.

O! Fili chare, noli nimis alte volare
Si nimis alte volas, poteris comburere pennas.

Advice to the would-be-wise:

Why do you insist on remaining ignorant, instead of Knowing Thyself?

MERCURIUS DE MERCURIO

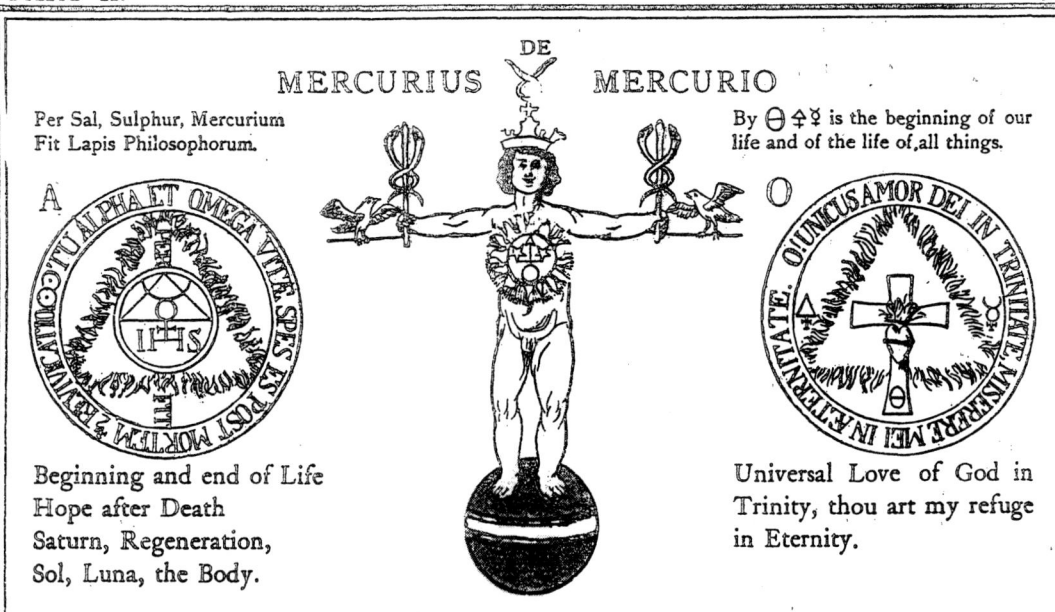

Per Sal, Sulphur, Mercurium
Fit Lapis Philosophorum.

By ☉ ♃ ☿ is the beginning of our life and of the life of all things.

A

Beginning and end of Life
Hope after Death
Saturn, Regeneration,
Sol, Luna, the Body.

O

Universal Love of God in Trinity, thou art my refuge in Eternity.

The Dew of Heaven and the Oiliness of the Earth are the materials for our work. It is therefore neither a mineral nor a metal. The pythagorean ¥ indicates that there are two mercurial substances in one root, Fire and Water, Ischschamaim, namely ☿ extracted from the substance, in which all metals are contained. It is a ☉ Dew of Heaven, but a metallic Dew, containing all colors. This Dew can be coagulated by the hermetic art and produces a sweet Salt, or Manna. Its father is the Sun, its mother the Moon, and from these two it receives its Life, Light, and Brightness. From the Sun it receives its Fire, from the Moon its Light. We find this dew in a coagulated state and also dissolved. It falls into the depths of the Earth, and its substance is the most subtle and ethereal part of the Earth; from above comes its soul and spirit, fire and light and enters the body of Salt. Thus it receives the powers of all things from above and from below. This mineral dew appears to us in colors of white, yellow, green, red and black. It appears corporeal to the external eye; but to the miners in the mountains it appears sometimes thick, watery and dripping. The best dew is the one which is coagulated like an Electrum or like transparent Amber. This heavenly Dew and its power is contained in everything. It is treated by the world with contempt and rejected by it. As it grows it becomes divided into two branches, white and red; both springing from one root ¥. This substance grows out of that one root, appearing like a white and red Rose of Jericho, and blooming like a Lily in the valley of Josaphat. It is often prematurely broken by the Miners and tortured by ignorant workmen. But the true Artist observes its influence by his developed internal senses, and gathers it, when it is ripe, with its flowers, seeds, root, trunk and branches. Let these hints be sufficient. It is neither a metal nor a mineral, but the mother of all metals and minerals, and their Prima Materia.

It is nothing else, but the coagulated Blood of the Red Lion, and the
Gluten of the White Eagle.

If you discover it, be silent and keep it sacred.

Trust to nobody but to God.

Experto crede Ruberto.

Fidelity has escaped from the Earth and is gone to Heaven. He has left
those men who cling to the Earth.

THE REVELATION OF JESUS CHRIST.

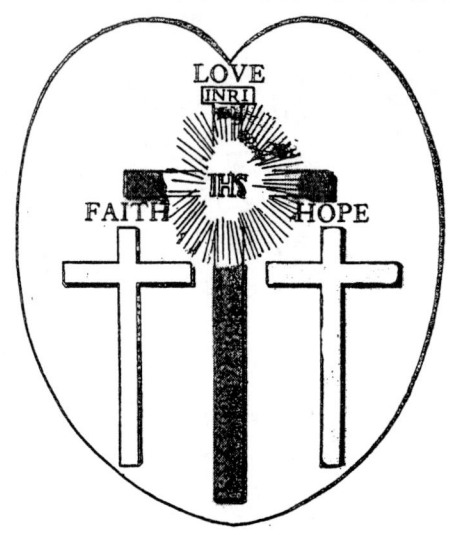

My Beloved Ones!

Every human being, in whom Jesus Christ the Son of God becomes manifest in and through the power of the Holy Ghost as the Son of the Eternal Father, is a true Christian; but it is evident to even a superficial observer, that until this day the spiritual power of the Christ-principle and the real living Christian Faith have never become fully manifest and revealed in humanity; for the Love of God, whose only true expression is the Universal Love of Humanity, made manifest in acts of Charity and Tolerance, is very seldom to be found among mankind, and in its place rule the Demons of Selfishness, Idolatry, Lust, Superstition, Skepticism, Injustice, Ambition, Intemperance, Envy, Greed, Luxury, Injustice, Theft, Murder and all the vices springing from Evil and Ignorance. All of this is entirely against the revelation of Christ, and is an obstacle in the way of His Regeneration. Why should we quarrel with each other? Is the Love of God, poured out over all mankind by the power of the Holy Ghost, exhausted? Has its Light become extinct? Have we become unreasonable animals? Does Heaven and Earth belong to only one class of men? Ah, no!

Oh, ye kings, knights and nobles pray to God that His Son Jesus Christ may become manifest in You! Oh! ye theologians, You should receive your knowledge from God and preach to the people the living Christ. You should do good for the love of Good, not for the love of money and remuneration. May you obtain the true knowledge of Christ, the son of God and the Virgin; may you be illuminated by Divine Wisdom and be a true Light to the world. Do not glory in your own powers, but glory in the power of God (*I. Cor. i., 31*). Then will dispute, damnation and heresy be ended, and One Love, the Love of God, will unite all. If all do the Will of God, they will all be of one Will, one Mind, one Faith and possess one Knowledge, one Happiness, in the Love of each other.

Oh, everybody! rich and poor ones, men and women, young and old, big ones and little ones, let us seek the revelation of Christ in each other's love, and we will all become peaceful, contented, patient and modest; we will then all belong to one church and be saved in and through the power of the love of Christ. Let no one pretend to love Christ, who does not love his brother, for God sees and judges the hearts of men. Love is the greatest power in the Universe, it is the fountain of Good and conquers all other powers.

Spiritual Love to God in Man is the source of all happiness; to experience it is divine and eternal Wisdom, Theosophia.

ROSICRUCIAN PRAYER

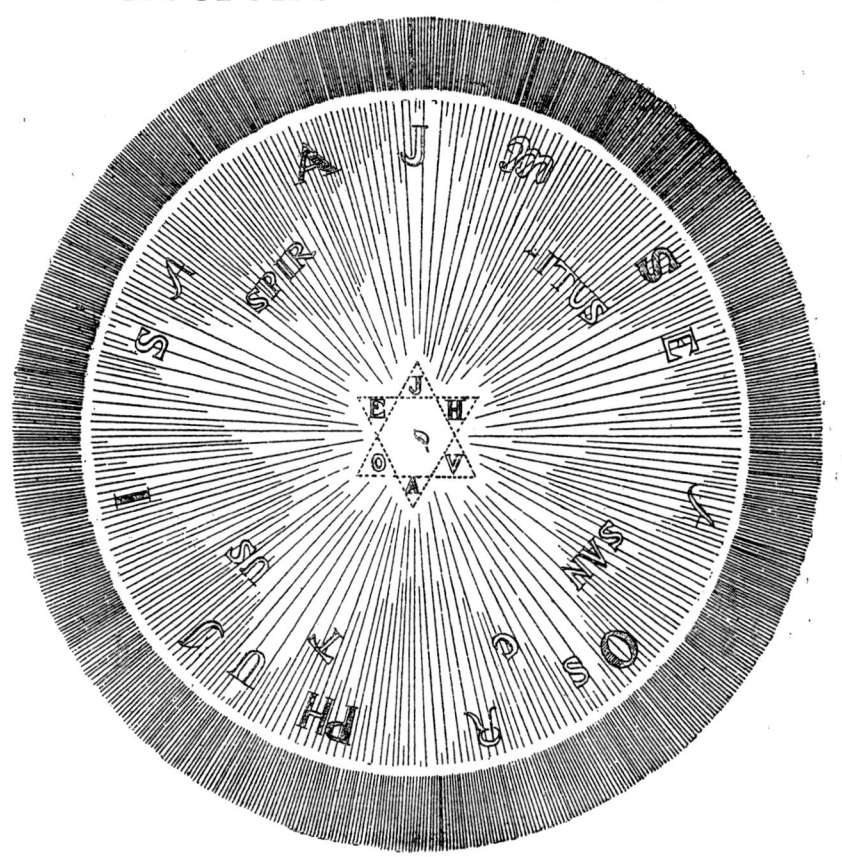

Eternal and Universal Fountain of Love, Wisdom, and Happiness; Nature is the book in which thy character is written, and no one can read it, unless he has been in thy school. Therefore our eyes are directed upon Thee, as the eyes of the servants are directed upon the hands of their masters and mistresses, from whom they receive their gifts. Oh thou Lord of Kings, who should not praise thee unceasingly and forever with his whole heart? for everything in the Universe comes from thee, out of thee, belongs to thee and must again return to thee. Everything that exists will ultimately re-enter thy Love or thy Anger, thy Light or thy Fire, and everything, whether good or evil, must serve to your glorification. Thou alone art the Lord, for thy Will is the fountain of all powers that exist in the Universe; none can escape thee. Thou art the helper of the poor, the modest and virtuous. Thou art the King of the world, thy residence is in Heaven and in the sanctuary of the heart of the virtuous.

Universal God, One Life, One Light, One Power, thou All in All, beyond expression and beyond conception! O Nature! Thou something from nothing, thou symbol of Wisdom! In myself I am nothing, in thee I am I. I live in thy I made of nothing, live thou in me, and bring me out of the region of self into the eternal Light. Amen.

ETERNITY.

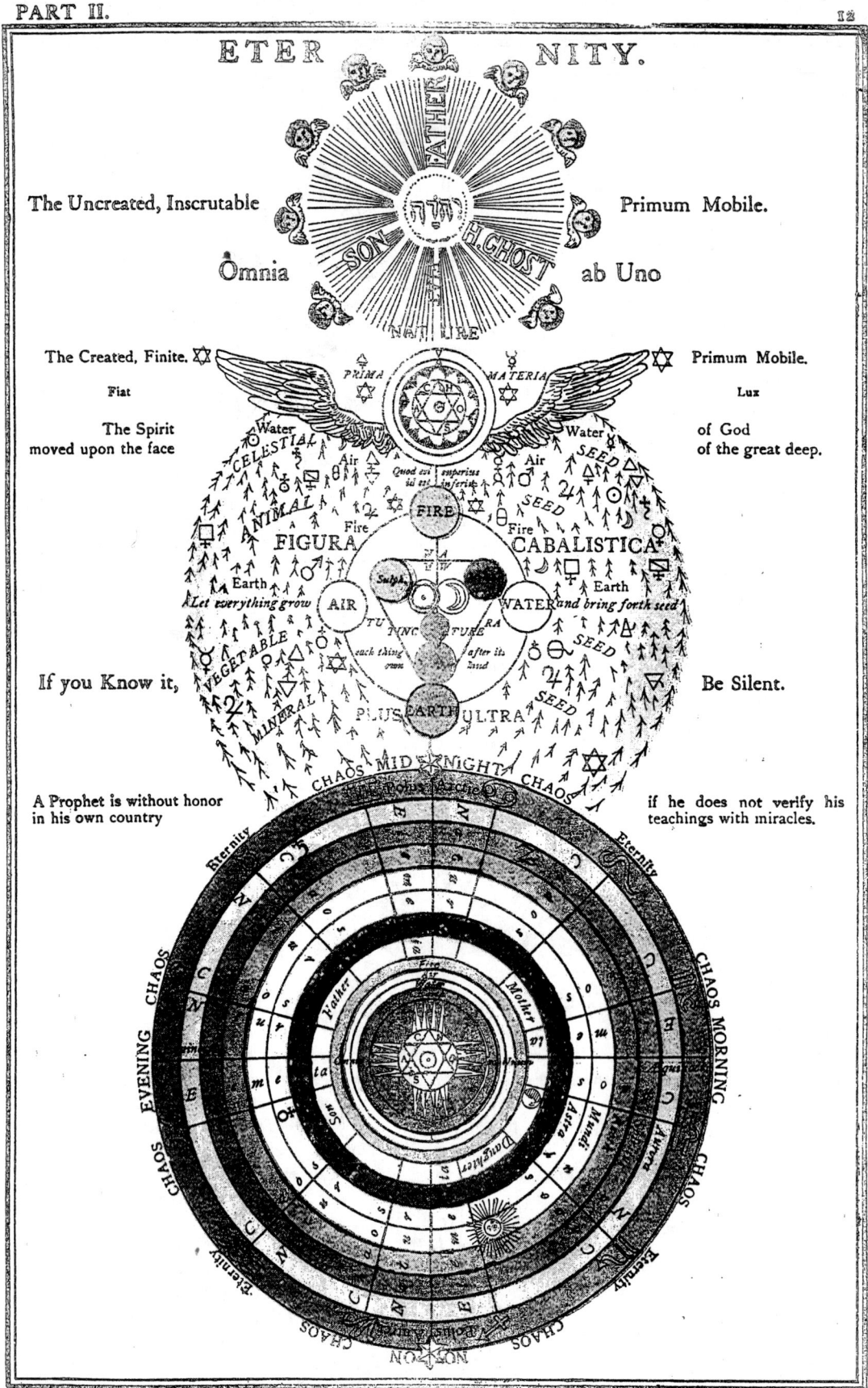

The Uncreated, Inscrutable — **Primum Mobile.**

Omnia — **ab Uno**

The Created, Finite. — **Primum Mobile.**

Fiat — **Lux**

The Spirit moved upon the face — **of God of the great deep.**

If you Know it, — **Be Silent.**

A Prophet is without honor in his own country — if he does not verify his teachings with miracles.